Smartcultures

**Nature's tiny Genius – Algae –
Reverses Environmental Pollution
and Regenerates Ecosystems**

*A green algae strategy that enables growers to
leave our fields better than we found them.*

Mark R. Edwards

THE GREEN ALGAE STRATEGY SERIES

www.AlgaeAlliance.com
www.AlgaeCompetition.com

Key words.

Food	Biofertilizer	Sustainability	Fertilizer
Water	Nutrient recovery	Ecosystems	Hunger
Regenerative	Organic farming	Smartcultures	Poverty
Bioavailable	Climate change	Energy	Drought
Algae	Micronutrients	Environment	Soil fertility
Soil crust	Global awareness	Algaculture	Malnutrition
Genetics	Renewable energy	Biotechnology	Pollution
Microalgae	Industrial farming	Ecology	Wetlands

ISBN 1456524690

ISBN 13 9781456524692

Smartcultures is the sixth book in THE GREEN ALGAE STRATEGY SERIES. Other titles are listed on page 193. They are available for free color PDF download at www.AlgaeAlliance.com for use by faculty and students in educational institutions. They are also available on Amazon.com and other retailers.

Dedication

To Ann Ewen, who lights up our garden and my green path.

To Sarah Edwards, who finishes grace before meals with, "Please God, bless this food and help people who don't have food get some."

To Blaine Metting, whose insightful but ignored research on algal nutrient-delivery systems and soil conditioners in the 1970s and 80s laid the foundation for smartcultures.

Contents

Smartcultures – Nature's Tiny Genius.........................vii

Preface ix

Chapter 1. Why Smartcultures?............................... 1

Chapter 2. The Stuff of Life – Soils 21

Chapter 3. What are the Forms of Agriculture? 35

Chapter 4. Microalgae – Nature's Nanobiofactories.............. 65

Chapter 5. Global Climate Change and Crops 73

Chapter 6. Mass Extinction of Fossil Resources...... 89

Chapter 7. Nutrient Recovery with ZooPoo360..... 115

Chapter 8. How Does Nutrient Delivery Work?..... 135

Chapter 9. Forms of Fertilizer 147

Chapter 10. How should we Address World Hunger? 165

Appendix I. Soil Macro and Microorganisms 183

Appendix II. The Use of Elements in Plants........... 187

Acknowledgements... 191

The Green Algae Strategy Series 193

Great Green Reading 194

Smartcultures – Nature's Tiny Genius

I am a farmer, a gardener and a lover of plants. Our organic garden flourishes with Mediterranean herbs, flowers, fruits, figs, vegetables and lots of garlic. I know my seeds will grow into plants that will:

- Require extensive husbandry to stay alive.
- Need constant weeding and tilling to grow to size.
- Consume expensive fertilizers and irrigation to thrive.
- Require diligent pest management to create plants of prize.
- Demand my constant vigilance, brain and labor to survive.

Our family and friends enjoy the beauty in, and bounty from, our garden. We rejoice in the visual panorama with its color, texture and aroma, as well as the birds, bees and butterflies. Yet we dream of self-regenerative plants – plants smart enough to flourish on their own.

Who would ever have imagined, that my husbandry has been amplified, all along, by tiny microflora in the roots of my plants; microorganisms that are ingeniously self-regulating and self-regenerative. These smartcultures that provide the foundation and food that enable my plants to flourish are:

- Nano-organisms, far smaller than I can see.
- Microbial communities, growing in the roots of plants and trees.
- Organisms working harder, faster and wiser than a billion bees.
- Workers that perform brilliant engineering, yet charge no fee.
- Nano-biofertilizer manufacturers that produce for free.

Who would have ever guessed – tiny plants smarter than me!

Preface

We are running really hard in a race to see which of our self-imposed maladies threatens our survival or the lives of our children first: nutritional deficiencies, affordable good food, natural resource extinction or climate chaos. Another stealth predator, ecosystem degradation and pollution, may strike many of us first. Smartcultures offers practical solutions for each of our survival challenges.

In the 11,000 years humans have practiced agriculture, communities survived or starved based on their relationship with their soil. Preserving fertile soils enabled farmers to grow crops to feed their families, communities and societies. Some of the earliest written documents on record are agricultural manuals that organized, preserved and conveyed soil and crop production knowledge. Indigenous civilizations, both ancient and current, worshipped the soil as the foundry of life. Why not venerate soil when the Latin name for man, homo, is derived from humus, the stuff of life in the soil?

Human societies cannot stand on an extractive, wasteful and pollutive industry to supply food. Very soon, the fossil natural resources on which our fossil food depends will no longer be available or affordable. Our current food production methods use precious non-renewable resources once, and then they are lost to the field. Modern agriculture wastes roughly half the fossil resources applied with each harvest and most the rest erode from fields and pollute our air, soil and water. Systemic extraction and erosion denudes soil fertility while massive pollution causes severe damage to human and animal health.

Soil ecosystems provide the foundation of human life because food production depends on fertile land to grow the produce that gives us energy, health and vitality. The intensity and scale of modern cropland use and abuse suggest we have much yet to learn and implement to sustainably manage the soil beneath our feet. Soils remain possibly the least understood of nature's critical ecosystems and certainly are among the most degraded.

Smartcultures

The survival of our children depends on gaining new knowledge and respect for soils. We need immediate actions that preserve and protect our croplands. The goal for smartcultures is to reverse modern extractive fossil food production and leave every field and the community ecosystem better than we found it. The smartcultures design engages farmers to transform our food production system from an extractive and pollutive industry to regenerative and clean, with more social equity. The new model also promises healthier foods that promote more valuable nutrients, vitamins and minerals per bite.

Smartcultures leverage nature's oldest energy production and storage method – photosynthetic algae – to support modern food production in the fields. Smartcultures transform our expensive, extractive and pollutive food production industry to a less costly, more productive system that regenerates soils while cleaning polluted air and water. Smartcultures enable farmers to leave every field and community ecosystem better than they found it.

Our world needs a reliable and affordable food supply that provides sustainable crops in the face of global climate change and the impending extinction of fossil-based resources – fertile soils, fresh water, fuels, fertilizers and agricultural chemicals.

Failure to develop a sustainable food supply will lead to a catastrophic food cascade where food literally falls off the shelf. A food cascade operates like a bank run where people buy, borrow, loot and hoard food based on anxiety about food insecurity.[1] A food cascade would create severe economic, environmental and social disruptions and very likely lead to famine, wars and mass starvation. Our modern, extractive, fossil foods put us a tiny step away from a food cascade.

Naysayers argue that current food surpluses will continue indefinitely. They may overlook the fact that over 40 countries experienced severe food riots in 2008 as food prices doubled and tripled.[2] They also may be unaware that our food surpluses are rapidly diminishing rapidly. Each decade we produce less food grains. Globally, growers lack the natural resources to substantially increase food production.

Our food supply is highly inelastic. Small surpluses cause prices to plunge while minor shortages, or the anticipation of shortages, produce abrupt price spikes.

When the world economic meltdown in 2008 sent shock waves around the world, governments responded by printing money. Consider that commodity markets today are leveraged on fossil resources substantially higher than the levels 2008 capital markets were leveraged on dollars. When a food cascade hits, governments will be unprepared to provide food. Global food stores are far insufficient. The meager food supplies will be quickly bought by the wealthy, leaving billions very, very hungry. Governments can print money in minutes, but food production takes years – assuming sufficient knowledge, labor and key natural resources are available – and the weather holds.

The food supply will crash, probably by 2031 in select food growing regions due to the unintended consequences of industrial agriculture. Each year, more countries become net food importers because domestic inputs for growing food have been consumed. Modern agriculture erodes the foundation of food production and has already robbed much of the world's best soils, fossil fuels and mined elements such as phosphorus, copper and zinc. When the fossil resources needed to operate industrial agriculture's heavy machinery become unavailable or unaffordable to farmers, food production will stop.

Modern fossil food production is non-sustainable because it:

- Extracts and over-consumes massive quantities of nonrenewable fossil fuel, inorganic fertilizers and agricultural chemicals mined from the earth's crust, uses them once and then allows them to flow from the field or be flushed into waterways. These chemicals poison marine life and create enormous dead zones.

- Over-consumes fresh water, including water from fossil aquifers that is not replaceable with normal rains.

- Systemically extracts soil nutrients and organics without full replacement.

Smartcultures

- Decimates the critical microflora – algae, bacteria, fungi and other microorganisms – that provide structure to soil and nutrition to crops.
- Severely pollutes and poisons air, soil and water with millions of tons of mined agricultural chemicals.

Modern agriculture has replaced biodiversity with monocultures that lack resiliency and place our food supply in severe jeopardy from global climate change and pest vectors. Most modern crops cannot tolerate temperature spikes, drought, fierce storms, dry winds and wild fires that occur with global warming. Genetically modified crops are not a solution because they multiply natural resource consumption and are exceptionally vulnerable to pest invasions.

Extinction ecology refers to something that ceases to exist locally in the chosen area of study but still exists elsewhere. This phenomenon, also known as extirpation, threatens industrial agriculture. The serious danger missing from text books is that when just one of the 21 critical fossil resources becomes unavailable, crops stunt or fail. We need a regenerative rather than extractive food production system less dependent on energy and external inputs.

> Sustainable MicroAlgae Regenerative Technologies – smartcultures – reengineer our food production system, beginning at its foundation – soil – with tiny microflora in plant roots that are ingeniously self regulating and self regenerative.

Smartcultures transform agricultural methods. Which makes more sense, mining and paying high prices for chemical fertilizers and using them once or recycling nutrients from the farm's waste stream? Rather than systemically extracting soil nutrients and organics, why not add nutrients and organics to soils? Instead of using chemicals that destroy soil microbes and soil structure, why not cultivate microbial communities that improve soil structure? Why continually degrade soil, promote erosion and pollute air, soils and water when growers can improve soil structure and reduce nutrient waste, erosion and pollution?

Smartcultures represent a set of technologies that mimic nature to provide enhanced foundation (soil structure) and food (nutrients) to plants. Every farmer and gardener knows plants thrive in amended soils; they grow faster, stronger and larger. Better soils enhance taste and texture. Smartcultures enable farmers to minimize the high cost of fertilizers by recovering nutrients from the farm waste stream. Target nutrients carried in algal biofertilizers can be delivered in precise amounts at specific times during a crop's growing cycle.

World hunger cannot be solved using the current methods of industrial or organic agriculture because both are too consumptive. Smartcultures integrate best practices from both forms of food production, with an emphasis on nature's way; organic techniques. The line between industrial and organic producers blurs with smartcultures because the methods are technology neutral. Farmers use highly productive combinations of chemical and organic fertilizers as well as high-yield seeds that best fit their growing conditions. Smartcultures will help all growers attain higher yields and better nutrition with fewer fossil inputs. Growers can save money and energy, reduce pollution and add rather than extract from their fields.

Microflora serve plants and animals at the bottom of the food web and remain the most undeveloped biological system on Earth. Nature places millions of microorganisms in the roots of every plant. They act similarly to microbial communities in the human gut to support natural processes that improve health and vitality. Smartcultures simply cultivate and amplify nature's nanocultures to improve food production and produce quality.

Even though each smartcultures element has been built and tested, the full integration of smartculture technology has not yet occurred. The recovery of energy and nutrients from waste streams has been prototyped but not implemented on a large scale. Solutions to nutrient delivery, crop productivity, soil conditioning and erosion are still under development. Current systems improve soils and crops, but only by 20—40%. We must do better and we can with collaboration. The recovery and reuse of waste-stream nutrients can solve critical fertilizer availability and affordability problems but the process needs extensive R3D – research, development, demonstration and diffusion. Much of this new knowledge base needed to successfully apply nature's nanotechnology has yet to be written. Of the millions of articles on agriculture, soils, fertilizers, energy and pollution, only a few pages have been written on the symbiosis of soil microflora and land plants. Except for the excellent work of Blaine Metting at the Pacific National Research Laboratory in Richland, Washington, and the considerable work by scientists in India and China on biofertilizers that fix nitrogen for rice production, the literature on microflora's multiple benefits to land plants is still largely an empty set.

Additional minds are needed to extend smartcultures R3D, broaden applications and discover new solutions. *Smartcultures* aims to stimulate communities of collaboration that will act quickly to avoid the mass extinction of fossil resources and to create food security for every human being on Earth.

Critics may argue that insufficient research exists to move forward with smartcultures. However, about half of our projected global population of 8.3 billion people by 2031 will live or starve based on our ability to change course quickly and produce sustainable and affordable food. Our current consumptive, expensive and wasteful food production practices are unsustainable. We must find ways to conserve our limited natural resources and to grow crops strong enough to flourish despite the vagaries of climate change.

We have only 20 years to make this course correction before fossil resources become extinct and food production crashes. The clock is ticking. What other options do we have? Let's work together to leave our fields and gardens better than we found them.

Mark R. Edwards
January 2011

Chapter 1. Why Smartcultures?

We know more about the movement of celestial bodies
than about the soil underfoot.   **- Leonardo da Vinci**

Since Leonardo da Vinci's time, we have learned about soils but we have failed to put enough of what we know into practice. In our hubris, we diminish soils as dirty dirt and treat soils like the carrier pigeon; disposable. Our society values distant bodies more than the earth beneath our feet because every year we spend over 100 times more on space flight than soils research.

Our behavior shows how we devalue soil. Globally we have abandoned over 30% of our fertile cropland in the past 40 years because we did not take care of them.[3] Much of the remaining cropland is so degraded that it takes twice as much fertilizer and water to achieve a stable yield. We need to listen to the sage advice from President Franklin Delano Roosevelt:

The nation that destroys its soil, destroys itself.

If a foreign country inflicted the current level of soil degradation and pollution of our precious ecosystems that we impose on ourselves, our nation would declare war on that country. As we deplete our soil, we silently erode the foundation of our society. When our soils and natural resources are no longer able to support food production, our society as we know it will end. Failing sufficient resources for food production, we will have no economy, no viable society and no national security.

Smartcultures

For the last 60 years, modern industrial agriculture has treated cropland as a disposable commodity. Systemic soil use, abuse and abandonment are foolish and wasteful. Much of the best cropland has been farmed for years and millions of those fertile acres will be unavailable to our next generation due to degradation and pollution. Cropland expansion requires more of every farming input because the new ground is less flat and fertile. Unfortunately, not only will our next generation be short of fertile cropland but they will also find the critical inputs for industrial food production gone.

Our food supply system needs to be reengineered so that agriculture produces sustainable and affordable food while minimizing or eliminating its current addiction to and massive extraction of fossil resources. Modern food production continuously extracts soil organics and nutrients while disrupting, depleting and compacting soils. Chemical fertilizers and pesticides destroy beneficial microflora communities that provide natural compounds to crops.

Modern agriculture is not sustainable because there simply are not enough fossil resources in the Earth's crust to meet its consumptive needs. Even if there were enough natural resources, industrial agriculture would fail because it disrupts, degrades and erodes its own foundation – soils and ecosystems.

Industrial agriculture over-consumes and then wastes the precious natural resources that food crops need. Farmers will soon find valuable resource reserves gone and prices far beyond their reach. In developing countries, fossil food inputs such as fresh water, fuels and fertilizers are already unavailable or unaffordable to many farmers. In addition, society cannot afford the health impacts and the severe environmental degradation caused by today's agricultural pollution.

Smartcultures

Smartcultures leverage the power of nature's tiny biofertilizer manufacturers. These microbial communities build stronger soils by gathering and delivering nutrients. They grow rich organic material that produces healthier and hardier plants.

Why Smartcultures?

Smartcultures enable farmers to recover and recycle nutrients lost in their waste streams and to use nano-sized algae on their fields. Microalgae provide an organic nutrient-delivery system for crops. Algal biofertilizer is loaded with the full spectrum of essential plant nutrients, stimulates natural plant growth hormones, and acts biologically to condition soils. Plants germinate and grow faster and hardier with immediately bioavailable nutrients. Stronger plants are better able to withstand the stressors of heat, drought and pests.

Microalgae make up the bottom of the food web and provide nutrients for higher-level plants and animals. Algae represent nature's oldest energy production and storage system. They attract vibrant communities of microflora that are beneficial to plants. Algal communities grow quickly and can restore soils that have been diminished by industrial agriculture.

Smartcultures add nutrient recovery and microalgae biofertilizer delivery to the actions associated with organic farming. Success in developing a sustainable food supply will require the cooperation and integration of industrial, organic and smartcultures farming methods. Fortunately, these diverse farming methods are compatible and smartcultures can help industrial farmers adopt sustainable, ecofriendly practices.

Smartcultures address several critical food-production challenges:

- **Nutrient erosion.** Smartcultures reverse the erosion of vital food micronutrients that are required for desirable produce taste, color and texture and for human health.

- **Nutrient extinction.** Smartcultures recover from local waste streams the full spectrum of macro- and micronutrients that plants need, reusing them to save money and conserve precious fossil resources.

- **Nutrient delivery.** Smartcultures deliver targeted, immediately bioavailable nutrients to plants, improving crop growth and development.

- **Soil conditioning.** Smartcultures rebuild soil – the foundation for crops – by adding organics and nutrients and improving soil structure, which increases crop health and vitality.

Smartcultures enable farmers to leave each field better than they found it in terms of *in situ* nutrients, organic material, soil structure and erosion resistance.

> *Smartcultures enhance crop production sustainably by leveraging Nature's nanotechnology to create nutrient and organic capital in soils to feed plants and build soil structure for stronger plants. Smartcultures recover, and recycle energy and nutrients from waste streams while minimizing pollution and reducing fossil resource consumption.*

Farmers may use smartcultures regardless geography, as long as their farms receive sufficient light to grow crops. Smartcultures are relatively easy to use and require only a few minutes per day of a farmer's time during the growing season. In most cases, smartculture technologies require no heavy lifting, operation of large equipment or exposure to agricultural chemicals or poisons.

Farmers who farm with smartcultures:

1. **Recover lost nutrients.** Smartcultures enable farmers to transform a cost to profit center by using algae to recover the reusable constituents of wastes instead of paying to have them incinerated, buried or transported to landfills. Some farmers may burn their biological wastes in a gasification or pyrolysis unit to create a low-cost carbon source for algae, and biochar, an excellent slow-release fertilizer and soil amendment.

2. **Grow algae locally.** Farmers who employ smartcultures may grow algae close to their fields, using nutrients recovered from the waste stream or other sources. Farmer can grow and flow the algae into irrigation, enabling the water to deliver targeted bioavailable nutrients to the crop. The algae applied to fields will continue to grow, as long as the soil retains moisture. Algae attract a broad spectrum of microflora beneficial to the crops

that build soil nutrients and organic capital. Algae can improve structure and unbind nutrients such a phosphorus and zinc that are locked in the soils.

3. **Act as land stewards.** Farmers who use smartcultures maximize the sustainability of food-production practices by using the Native American model of land stewardship that looks forward seven generations. Smartcultures stewardship of the land may include:

   - Organic production that maximizes the use of organic fertilizers (from living organisms) and minimizes the use of agricultural chemicals such as inorganic (mined) fertilizers, pesticides and other poisons.

   - Water management designed to minimize water use such as micro-drip irrigation.

   - No-till farming, in which no or minimal cultivation disturbs and disrupts the soil.

   - Nutrient management and soil structure management that minimizes erosion and pollution.

   - Crop rotation and the use of cover crops to sustain soil nutrients and organics and minimize erosion.

   - Biological pest control to minimize agricultural poisons.

Preliminary experience shows that smartcultures can increase farmers' income by improving the quality and quantity of crops 20–40%. Farmers can save money and energy by lowering their consumption of fossil fuels by 30–50% and reducing the need to apply chemical fertilizers by 30–50%. Smartcultures also decrease air, soil and water pollution by 80 – 90%.

Productivity, resource consumption and cost savings vary substantially based on soil fertility and structure. Smartcultures provide the highest added value for abandoned cropland or compacted soils depleted of nutrients and organics that are prone to erosion. However, every field and every crop can be enhanced with the smartculture methods.

The Root of the Problem

The development of roots was a substantial evolutionary compromise that was necessary for land plants. Roots were necessary to hold land plants in place, as well as to create a plumbing system to extract water and nutrients from the soil. Unfortunately, roots created a heavy production drag for plants because growing and maintaining root structures consumes about 30% of a plant's energy. Roots also anchor the plant in place, creating a dependence on soil moisture and bioavailable nutrients present in the plant's root zone or rhizosphere. The rhizosphere is the narrow region of soil that is directly influenced by soil microorganisms where roots can absorb them.

Communities of nano-sized algae have lived symbiotically with plants since land plants evolved from algae 500 million years ago. As land plants moved inland from ancient shorelines, they needed foundation and food but they had no roots. Algae formed soil crusts that provided the foundation that enabled plants to withstand wind and weather. Algae also supplied food energy in bioavailable nutrients before plants had roots. Thus, algae provided the bridge that enabled water-based plants to adapt to terrestrial ecosystems.[4] Today, algae continue to live symbiotically with land plants as algae, and the microflora they attract, provide plants with a full set of macro- and micronutrients while they continuously improve soil structure.

Roots significantly limit plant growth because these delicate appendages can take up nutrients only when they are in a bioavailable form, usually after they have been broken down by soil microbes. Therefore, even though a nutrient such as phosphorus (P) may be present in soil, a plant cannot use that nutrient until it has been processed by microorganisms into a digestible form called reactive or bioavailable P. When a plant experiences a growth phase without a needed macro- or micronutrient in bioavailable form, growth typically occurs, but at a suboptimal rate. Nutrient limitation:

- Stunts or stops plant growth and development.

- Decreases total biomass and produces skinny, "weathered" or wilted plants.

- Reduces or halts fruiting (seeds and nuts are the fruit of the vine).

- Diminishes appearance, color, shape, size, nutritional value, texture and taste.

Modern farmers use chemistry to address nutrient limitation.

Chemical Fertilizers

Modern farmers use soil chemotherapy to replace soil nutrients as they apply inorganic (mined, chemical) fertilizers to soils. Synthetic N (nitrogen) fixation enabled a 20 fold reduction in the volume and weight of fertilizer relative to manures, drastically reducing fertilizer transport and application costs. Fertilizers are leveraged on fossil fuels because manufacturing the fertilizer needed to grow one acre of corn consumes the energy equivalent of 30 gallons of gasoline.[5] Currently, the total cost of fertilizer is cheaper and easier than organic production because fossil fuels from which those fertilizers are made or mined are relatively cheap. In many food growing regions, including the U.S. Midwest, less than 30% of the manure or compost needed to sufficiently fertilize fields organically is available.

Farmers know that non-organic fertilizers may not be bioavailable to plants and that soils may or may not have sufficient soil microbes, since agricultural chemicals typically kill them. Consequently, chemical fertilizers are massively over-applied, with the result that the nutrients in the fertilizers erode or change to other molecular forms that are not bioavailable to the crop.

For example, P (phosphorus) acts to stimulate early growth and strong root formation. The P in synthetic chemical fertilizer quickly bonds with other compounds in the soil, putting it into a form that plants cannot access. Bioavailable P – in a form that plants can access – may constitute less than 0.1% of the total P present in the soil.[6] In addition, the form of P in inorganic fertilizer is extremely soluble and easily rinsed from the field, which is why P is one of the most common pollutants of our waterways and ground water.

Smartcultures

Modern farming causes plants to sub-optimize growth and development by failing to provide sufficient bioavailable nutrition in the soil at critical times in the growth cycle. Nutrient deficiency, especially a lack of micronutrients is the primary reason that much modern produce such as field tomatoes have so little flavor. The nutrient deficiency problem is amplified by another consequence of modern agriculture – compacted soil that constrains root growth. Smartcultures solve both the nutrient insufficiency and soil compaction challenges by creating a dynamic delivery system for organic plant nutrients that also conditions soil, enabling plants to grow root systems that are 10-100 times bigger and deeper.

Plants need 17 elements for healthy growth and development. They accept no substitutes for these elements, which they may draw from natural organic or synthetic, inorganic sources. Inorganic nutrients must be broken down by microbes in the soil, typically algae, before they can be absorbed by plants. Seeds may germinate and plants may grow with insufficient nutrients, but they do not prosper. In fact, plants need each of the 17 elements in Table 1.1 to set fruit, which is the ultimate objective in growing food crops.

Nutrient need is based on the weight of the dry matter of the plant. Macronutrient requirements are roughly 1000 mg (or more) per kilogram of plant dry matter. Micronutrients are required in smaller amounts, typically less than 100 mg per kilogram of dry matter. As plants grow, they remove these nutrients from the soil. Farmers often fail to replace the micronutrients because they are not readily available in most commercial fertilizers.

Food crops grown in nutrient-deficient soil contain up to 75% less of the deficient nutrient than those grown in fertile soil.[7] With the exceptions of carbon, oxygen and nitrogen in nitrogen-fixing legumes, plants cannot pull nutrients from the air. Plants are limited to the building blocks in the bioavailable nutrients dissolved in the soil moisture found in their root structure.

Table 1.1 Required Nutrients for Crop Production*

Macronutrients		Micronutrients	
Nitrogen	(N)	Boron	(B)
Phosphorus	(P)	Copper	(Cu)
Potassium	(K)	Chlorine	(Cl)
Carbon	(C)	Iron	(Fe)
Oxygen	(O)	Molybdenum	(Mo)
Hydrogen	(H)	Manganese	(Mn)
Sulfur	(S)	Nickel	(Ni)
Magnesium	(Mg)	Zinc	(Zn)
Calcium	(Ca)		

* See Appendix I for a table that shows the metabolic functions of each crop nutrient listed in Table 1.1.

Erosion of Nutrient Density

The post-World War I Green Agricultural Revolution increased crop yields but half our global population, 3.5 billion people suffer from caloric and micronutrient deficiencies.[8] In wealthy countries, many people consume thousands of "empty calories," yet remain malnourished because their diet is deficient in essential vitamins, minerals and trace elements. Calories are empty in the sense that the foods have high levels of sugar, salt and starch, but deliver few nutrients per calorie consumed. Consumers want to consume foods that deliver more nutrients per calorie rather than less. In *The End of Food,* Paul Roberts describes how and why nutrient erosion degrades the sensory qualities of food, especially color, texture and taste.[9]

Industrial agriculture's focus on increasing yields has caused slow, yet systemic erosion in the nutritional quality of our food. The concentration and range of essential nutrients in the food supply has declined in each of the last few decades, with double-digit percentage declines in iron, zinc, calcium, selenium and other nutrients essential for human health.[10] Consequently, each calorie delivers more sugar and starch with fewer vitamins, minerals and trace elements. The erosion of the nutritional value of produce enables low cost foods that shift consumer costs from food to health. Empty calorie foods contribute substantially to obesity, malnutrition and diabetes.

Breeding plants to grow bigger produce causes the plants to devote energy in head or seed production at the expense of deep roots. Industrial agriculture makes deep roots unnecessary because large amounts of applied chemical macronutrients (but not micronutrients) are readily available in the topsoil. Modern fertilizers act like junk food for plants – blimping produce with extra size and water.

Industrial farms produce high quantities of cheap foods that are high-fat and composed of refined grains, added sugars, salt and fats. Tom Vilsack, the U.S. Department of Agriculture (USDA) Secretary stated that his administration would put "nutrition at the center of all food assistance programs" because obesity healthcare problems have reached epidemic levels. More than 65% of U.S. adults are overweight or obese. In 2008, medical spending on obesity-related conditions cost consumers $147 billion, which represents about 10% of all medical spending. Annual medical costs have nearly doubled in the past decade from $78.5 billion in 1998. The typical obese American spends 42% more on healthcare or $1,429 more each year than the approximately $3,400 annually spent by normal-weight Americans. Obesity acts as a strong risk factor for Type 2` diabetes, heart disease, hypertension and stroke.

In *Death by Supermarket: The Fattening, Dumbing Down, and Poisoning of America*, Nancy Deville notes that four of the top 10 causes of death in America with chronic diseases linked to diet: heart disease, stroke, Type II diabetes and cancer.[11] The diets of the poor (and many of the affluent) result in obesity and a litany of health disorders. The Centers for Disease Control estimates that one in three American children born after 2000 will develop Type II diabetes. Modern agriculture has decreased the cost of food at the appalling expense of multiplying the cost of health care.

Scientists do not understand the nutritional benefits linked to the thousands of phytochemicals produced by plants. Since human

10

evolution involved eating natural foods from fruits and vegetables, the human body responds with greater vitality when more, rather than fewer, phytochemicals are present in produce. As farmers have pushed plants towards maximum yields, produce delivers more sugar and starch per serving and lower levels of health-promoting compounds such as vitamins, minerals and antioxidants.

Part of the nutrient density erosion occurs not due to available soil nutrients but to plant breeding. Breeding (hybridizing or genetically engineering) plants that can tolerate close spacing constrains root growth and multiplies the amount of fertilizer and water needed. Plants that lack deep roots are less stable and more prone to stresses such as heat, drought and pests. Shorter roots limit the plant's access to valuable micronutrients, which are the building blocks for health-promoting phytochemicals such as antioxidants.

Farmers accelerate plant growth with substantial added nitrogen (N) applied near the top of soil. Fast-growing grains, fruits and vegetables spurred with extra N display a "dilution effect." Plant cells are weaker, more watery and the available nutrients are diluted. Consumers buy larger and heavier produce but are paying for more water weight.

Modern crops become lazy and addicted to nitrogen-phosphorus-potassium (N-P-K) chemical fertilizers because inorganic fertilizers make nutrients so readily available. Why should plants waste energy growing deep roots when chemical fertilizers are at their feet? Consequently, crops ignore the harder-to-pull natural soil nutrients and organic matter in favor of the easy-to-reach chemical nutrients. Research with wheat, corn and vegetables show that modern high-yielding varieties generally have lower concentrations of nutrients than traditional, lower-yielding varieties.

The tradeoff between yield and nutrient level seems to be widespread across crops as plants partition their limited energy among different goals. In grains such as corn, wheat, rice and soybeans, higher harvest weight comes at a cost of diminished concentration of vitamin C, levels of lycopene (a critical antioxidant that makes tomatoes red), and beta-carotene (a vitamin A precursor).

The same nutrient erosion effect occurs in high-production dairy cows that produce milk lower in protein and other nutrition-enhancing components per gallon. High-production dairy cows, similar to highly productive crops, are substantially more vulnerable to a wide range of metabolic diseases, infections and reproductive problems.

In contrast to chemical fertilizers, organic fertilizers provide a balanced mix of nutrients, and they release the nutrients gradually, which promotes strong root development and longer roots that absorb more nutrients.[12] Similarly, reducing the use of pesticides boosts phytochemical production significantly.

Nutrient Extraction

An unintended consequence of growing denser stands of crops is the accelerated extraction of soil nutrients and loss of soil organics. A review of 110 years of yield data from the world's oldest continuous cotton experiment concluded that soils with more organic matter produce higher crop yields with better nutrient profiles.[13] Modern farmers have no practical way to replenish organic matter since composing organics takes at least a year and each ton requires ten tons of organic material.

Each harvest extracts about half of added soil nutrients and most the residual typically erodes from the field on the wind or water. Some nutrients percolate below the root zone and ends up polluting ground water. Industrial farmers typically replace the natural N-P-K fertilizers with chemical versions, but they do not replace the critical soil organics or micronutrients because they are generally not available in commercial fertilizers. For many large farms, sufficient compost is simply not locally available. Consequently, crops grow rapidly with the N-P-K fertilizers but, unseen by consumers, the resulting produce lacks the spectrum of valuable phytochemicals such as antioxidants.

Nutrient Extinction

In natural settings, fertilizers such as P are recycled roughly 50 times in plants and animals before flowing into the sea. Modern farmers apply newly mined and refined P to their fields once each growing season. Half the precious nutrients are removed with the harvest and much of the rest flow from the field in the agricultural waste streams. Since modern chemical fertilizers are inefficiently absorbed by plants because they are not in bioavailable form, farmers must apply far more fertilizer than the plants can actually use. Excess fertilizer contributes significantly to run-off and erosion that flow to wetlands, streams, lakes and groundwater.

Fertilizers add to greenhouse gases and atmospheric dust. Fertilized soils release more than two billion tons of greenhouse gases every year, especially CO_2, methane and nitric oxides. Each acre of corn production adds 2.25 tons of CO_2 to the air plus nitric oxide that has 296 times the warming capacity of CO_2 and also disrupts the protective ozone layer.[14] Nitric oxides take multiple forms and are often referred to as NO_x.

Smartcultures address each of these challenges with sustainable solutions. The core concept comes from Aristotle, who suggested using nature's way. Smartcultures mimic nature by employing nature's original nutrient-delivery system for plants – algae – to deliver targeted organic nutrients precisely when plants need them. Crop yields increase significantly because the plants receive targeted nutrients in a form immediately available to them, at each step of growth, development and fruiting. Plants grown with smartcultures have larger, stronger and deeper root systems, enabling them to absorb the micronutrients that give produce an enhanced nutritional profile with better size, color, taste and texture.

Bioavailable nutrients are nano-sized and are quickly absorbed by plants, minimizing or eliminating nutrient waste, erosion and pollution. The algae continue to grow in the field, building stored energy in humus that plants can tap at a later time. Algae attract symbiant communities of microflora that, together, create a rich soil crust that stores nutrients and minerals while minimizing wind and water erosion.

Industrial Agriculture versus Smart Agriculture

Table 1.2 considers how modern industrial agricultural technologies compare with smartcultures in meeting the current food production challenges.

Table 1.2 Comparison – Do Less and Do More.

Challenges	Industrial agriculture *DO LESS* to meet current challenges because they:	Smartcultures *DO MORE* to meet current challenges because they:
Food To create sustainable good food.	• Maximize crop production at the high cost of losing micronutrients, taste and texture. • Consume fossil resources at an unsustainable rate.	• Create high production while retaining or adding to micronutrients, taste and texture. • Consume fossil resources minimally.
Water To consume fresh water efficiently, in ways that will not lead to its extinction.	• Massively overuse fresh water for irrigation. • Waste about 50% of the water used, due to inefficient water delivery and poor soil structure. • Allow substantial runoff of water, due to compacted soils.	• Reduce total fresh water consumption. • Reduce water use through efficient irrigation delivery and improved soil structure. • Minimize runoff of water by improving soil structure.
Extraction To use the mined resources needed for agricultural chemicals in ways that keep them affordable	• Cause a net depletion of soil nutrients with each crop. • Cause a net depletion of soil organics with each crop. • Use mined inorganic fertilizers.	• Add soil nutrients with each crop, using microflora. • Add soil organics with each crop, using microbial flora. • Use home-grown organic fertilizer delivered by

and avoid their depletion.	• Use mined agricultural chemicals.	algae. • Minimize chemical use by stimulating natural plant compounds.
Energy To use less energy and use renewable energy.	• Rely heavily on diesel fuel for cultivation. • Incur huge energy costs to produce synthetic nitrogen fertilizer. • Incur high extractive costs for chemical fertilizers. • Incur high refining, packaging, transporting and application costs. • Incur high costs in manufacturing agricultural chemicals.	• Reduce the need for cultivation and diesel fuel. • Enable algae to fix N_2 near or on the plants in the field. • Recover, reclaim and reuse nutrients from waste streams. • Minimize fertilizer costs by growing fertilizers in algae near the fields. • Minimize chemical use by stimulating natural plant compounds.
Poisons To avoid pollution and killing beneficial soil microbes.	• Over-apply fertilizers. • Heavy application of pesticides, herbicides and fungicides. • Kill beneficial microbial communities.	• Use precision fertilization. • Stimulate plants' natural defense mechanisms and avoid poisons. • Enhance microbial communities in soils.
Pollution To avoid creating an expensive health and cost externality for society.	• Over-fertilize, causing pollution of the entire ecosystem. • Cause CO_2 and black soot pollution. • Cause the volatilization of high levels of nitric oxides (NO_x). • Enable severe wind, rain and irrigation erosion of degraded soils.	• Use precision fertilization, causing no or minimal pollution. • Use minimal diesel fuel to grow crops. • Cause minimal volatilization of nitric oxides. • Build a stronger soil structure that is resistant to wind and water erosion.

	• Cause massive dust pollution by cultivation.	• Minimize dust pollution with no-till farming.
Food To avoid production techniques that degrade the nutritional quality of food.	• Add nitrogen to stimulate fast growth. • Promote fast-growing plants with watery cells susceptible to heat and drought. • Erode nutrient density, especially of micronutrients. • Cause loss of good color, taste and texture.	• Supply organic nitrogen for quality growth. • Allow slow growth of plants to create strong cells resistant to stressors. • Grow produce packed with micronutrients, vitamins and minerals. • Retain good color, taste and texture.

Smartcultures align with organic production techniques where possible, yet represent a hybrid of methods. Smartcultures reverse several common agricultural practices in order to make food production sustainable and more compatible with human societies. Smartcultures require significantly less fresh water and fossil energy, which should lower food prices. Smartcultures reverse the practice of extracting nutrients and organic humus by building a rich foundation for food crops.

The practice of nutrient erosion must end so that plants can, once again, provide the full range of phytochemicals essential for human and animal health and vitality. We need to find natural ways to feed crops with sustainable and affordable fertilizers and to stimulate plants' natural defense mechanisms so that we can avoid poisoning our fields and waterways. We also need to rethink the ways we use and deliver fertilizer and agricultural chemicals to reduce costs, preserve supplies and avoid erosion and pollution.

Why Weren't Smartcultures Developed Earlier?

Nature created a special nutrient-delivery system for plants eons ago, and smartcultures simply intensify nature's excellent work. The core

Why Smartcultures?

Smartcultures methods are natural processes and are not patentable. Agribusiness corporations and their plethora of congressional lobbyists have created subsidies and incentives for farmers to use and then become addicted to expensive synthetic chemical compounds that are patentable and very, very profitable.

Farmers in China employed a preliminary form of smartcultures centuries ago. The Chinese have used blue-green algae that fixes N from the air to support rice production since at least the 11th century. Scientists in India rediscovered the value of algae biofertilizers in the 1930s and produced a considerable body of excellent work on what they call "algalization." Algalization focuses on biological mechanisms for fixing N in order to reduce farmers' dependence on nonrenewable resources and to avoid the cost of fossil fertilizers.

Organic farmers have used their ingenuity to gain many of the benefits smartcultures offer. They grow cover crops, apply organic fertilizers and compost to replace nutrients and build soil organics. To the degree possible, organic farmers avoid the use of fossil fertilizers and agricultural chemicals, which may improve food nutritional quality. Organic farmers also act as stewards of the land to maximize sustainable crop production.

Smartcultures build on centuries of experience with organic farming. The nutrient recovery process with algae parallels composting and adds a number of advantages, including speed, efficiency and far lower energy consumption. Targeted nutrient delivery parallels the application of organic fertilizer and adds several advantages, including fertilizer application efficiency and the ability to supply specific nutrients at the precise time plants need them.

Farmers practicing smartculture methods may find they need a combination of organic and chemical fertilizers to meet their crops' nutritional needs. They may adopt modern precision irrigation techniques to minimize water loss while maximizing the effectiveness of nutrient delivery. They also may use genetically engineered seeds that are drought- or heat-tolerant, or that are more efficient in their metabolism of key nutrients such as N and P.

Smartcultures

While smartcultures offer many benefits, several limitations apply.

Limitations

Smartcultures do not solve all the challenges facing farmers. Nutrient recovery systems that are being built on a commercial scale will need to be downsized for individual farms. Similarly, a variety of algal production systems are available for use near fields, but those systems need to be optimized for local production and the costs reduced. Targeted nutrient delivery works well on irrigated cropland, but additional research needs to extend applications to dry land farming. Smartcultures displace some, but not all, chemical fertilizers, unless the farm waste stream is large, such as a dairy or feedlot.

The technology makes plants hardier and better able to withstand the stresses of heat, drought and severe storms but probably contributes only about a 20% increase in robustness. Similarly, the process reduces, but does not eliminate, the need for fossil resource inputs for crop production, especially water and fossil fuels. While smartcultures substantially reduce pollution through the use of less fertilizer, agricultural chemicals and cultivation, some pollution still migrates to the external ecosystem.

Lessons learned by India's scientists from 1930s forward are informative:

- Farmers were reluctant to trust organisms they could not see.
- Biological fertilization works more slowly and creates less visual impact than chemical fertilizers.
- Knowledge transfer sometimes was poor between agricultural extension agents and farmers.
- The methods used to produce algal cultures consumed considerable cropland and were labor intensive.
- Poor quality control of algal cultures by third-party producers led to substantial variation in crop production, which negatively impacted farmers' interest in biofertilizers.
- Chemical fertilizers were still needed, especially to supply P, K, zinc and iron.[15]

Why Smartcultures?

India's experience provides a punch list of issues that need to be addressed in order to enable smartcultures to achieve their potential.

Smartcultures knowledge transfer and production systems with a three-year payback have become the goals for the smartcultures project. We want to ignite a debate about best practices for our food supply. We want to promote the adoption of food production methods that are less consumptive, more sustainable and produce healthier food with more nutrition. Most farmers interviewed for this smartcultures project reported that they would immediately employ algae to recover waste nutrients and for targeted nutrient delivery if they knew how, and if the process paid for itself in three years or less.

Farmers interviewed also want to be able to reduce their consumption and costs for water, fuels, fertilizers and agricultural chemicals. Every farmer wants a cost-effective means to improve soils, produce higher yields, use fewer external inputs and create less pollution. Farmers were clear that they would quickly adopt smartcultures methods if they could see successful operational farms, and if the nutrient recovery and delivery systems were economical and practical.

The lack of smartcultures demonstration facilities represents a significant limitation that can be overcome only with a build-out of systems. Adoption of smartcultures is limited by the lack of R&D, and lack of successful models for nutrient recovery with algae or vendors with algal nutrient-delivery systems. The USDA does not currently finance smartcultures research, although it supports a small amount (less than 1% of its research budget) of research on organic farming. The call to action here may prompt R&D funding and spur valuable research. Several companies are independently working on using algae to bioremediate wastewater, and those systems may be adapted to smartcultures. If algal nutrient-delivery systems work as well as preliminary research suggests, many vendors will soon be providing turn-key or leased systems to farmers.

Smartcultures offer farmers a green algae strategy to improve crops by enhancing soils. The next section focuses on the most important six inches on Earth; soils.

19

Chapter 2. The Stuff of Life – Soils

Land is not merely soil; it is a fountain of energy flowing through a circuit of soils, plants, and animals.

- **Aldo Leopold**, *A Sand County Almanac*, 1949

The energy for human life, our sustenance, comes from products of the soil. Fertility makes the major difference in produce quality, which is the "fountain of energy flowing through the circuit of soils, plants and animals." Soil fertility depends on maintaining sufficient nutrients, minerals, organic matter, moisture and microorganism communities.

Human societies depend on the knowledge and the actions of farmers to sustain fertile soils so they can provide food and fiber for current and future generations. Modern agriculture puts both current and future generations in jeopardy due to poor soil husbandry, especially the destruction of soil microbes from soil cultivation, compaction and agricultural chemicals.

The National Institutes of Health initiated a Human Microbiome Project in 2007 because health researchers believe the microscopic guests on and in each human body are really ancient allies. Scientists are concerned that medical practices are undermining the natural benefits from microflora by killing them in a manner similar to way modern farmers are destroying microbial communities in the soil.[16] The human body has about 10 trillion human cells, but there are 10 times more microbial cells such as archaea, algae, bacteria, viruses and fungi that provide many natural benefits. The number of microbial genera in and on plants is probably several times greater than those living in the human body. Soil microbes act in symbiosis with plants to provide a wide range of benefits.

> *To be a successful farmer one must first know the*
> *nature of the soil.*
> - **Xenophon**, Oeconomicus, 400 B.C.

Soils acts as a living system to promote plant growth through the provision of bioavailable nutrients to the plant roots. Soil organic matter retains water for nearly continuous plant uptake. Soils store greenhouse gasses and resist physical degradation from wind and water. Fertile soil maintains a network of interconnected pores that provide pathways of low resistance for root growth and provides room for microorganisms to flourish.[17]

Fertile soil is not an inert medium, but a mixture of water, air, minerals and organic matter. In most soils, minerals represent around 45% of the total volume, water and air about 25% each, and organic matter 2-5%.[18] The mineral portion consists of three distinct particle sizes classified as sand, silt or clay.

The magic of natural soil fertility comes from the combined work of earthworms, arthropods and the microbial community which was recognized by Nathaniel Hawthorn as the "Divine chemistry works in the subsoil."

Soil health depends on the organic component that houses many living creatures along with dead material in various stages of decomposition. An acre of living soil may contain 900 pounds of earthworms, 2400 pounds of fungi, 1500 pounds of bacteria, 133 pounds of protozoa, 890 pounds of arthropods and algae, and possibly some small mammals.[19] An acre of soil may contain over 10,000 species of microorganisms which contributes significantly to the biodiversity inorganic soil.[20] Appendix II provides a brief description of the primary soil macro and microorganisms.

Soil organic matter interacts to influence soil biological, chemical and physical properties and consists of:

- Raw plant residues and microorganisms (1-10%).
- Active organic traction (10-40%).
- Resistant or stable organic matter (40-60%), also called humus.[21]

People talk of agricultural practices that are differentiated by their input source – chemical (synthetic/industrial) or organic. From a plant's point of view, synthetic versus organic has no meaning because plants are blind to source. If a plant's roots find a needed bioavailable nutrient, plants absorb the nutrient and put it to use building structure or storing energy.

Microbial Communities

The size of the living community in the soil, the microbial biomass, is positively related to soil organic matter and typically varies between 1% and 5% of the organic component. Algae acts like honey to attract vibrant communities of microorganisms that serve plants as they act as agents for:

- A bioavailable source for macro and micronutrients.
- An immediate sink of C, O, N, P and S.
- Water mobility, transfer and retention.
- Nutrient transformation and pesticide degradation.
- Mineral weathering and soil formation.
- Stabilization of soil aggregates and soil structure.
- Production of microorganisms antagonistic to plant pathogens and parasitic nematodes.
- Production of plant growth regulators.[22]

Soil organisms–from the tiny algae to the large earthworms and insects–interact with one another in the rhizosphere as they create active communities that serve plants and each other in the soil ecosystem. Organisms not directly involved in decomposing plant wastes feed on each other or other substances they release. These microbial communities break down organic components and enrich soils by releasing biochemicals. These valuable building blocks can be directly absorbed by plants, but only if they are available in the soil and broken down into tiny molecules.

Roots stimulate soil microbes by releasing specific substances into the soil that serve as food and signals for select organisms. At various growth stages or when stressors such as pests threaten, plants release substances that stimulate specific populations of microorganisms capable of releasing the needed micronutrient.

Active soil organic components work together with microorganisms to bind small soil particles into larger aggregates. Aggregation is critical for good soil structure, nutrient cycling, aeration, water infiltration and resistance to erosion, clodding and crusting. The resistant or stable fraction of soil organic matter enhances nutrient holding capacity.[23] Improvements in soil physical structure facilitate easier tillage, which uses less energy, increases water storage capacity, and reduces erosion. Enhanced soil structure enables deeper and more diverse root systems for drawing nutrients. Improvements in nutrient cycling also reduce the need for additional fertilizers.[24]

Soil organics

Organic matter contains essential plant nutrients and is derived from plant residues (top growth and roots), plus residues from microflora in various stages of decomposition. Accumulated organic matter provides a storehouse of plant nutrients that releases nutrients in a plant-available form. Additional chemical fertilizer may increase crop yields but decreases organic matter because organic matter is lost with each crop and chemical fertilizers promote short roots, soil compaction and erosion.

The stable organic fraction, humus, is important for soil structure because it holds individual mineral particles together. Plants do not draw their nutrients directly from organic matter but from the mineral elements distributed in soil aggregates. Humus gives soil the dark brown or black color and contributes to fertility by providing mineral elements during its biodegradation. Humus aids in water filtration, movement and retention. Humus supports root strength and length, aids in soil erosion resistance and soil aggregations that hold minerals.[25]

Soil that contains 4% organic matter in the top six inches may hold 200,000 pounds of organic matter per acre.[26] This 100 tons of organic matter contains about 5.25% N, amounting to 4,200 pounds of N per acre. Assuming a 5% release rate each growing season, organic matter supplies about 210 pounds of N per acre to the crop.[27] If the organic matter degrades without replenishment, additional fertilizer is necessary to support crop yields.

Industrial agriculture uses chemotherapy to fields which increases yields in the short term but leaves fields exhausted in the long term. Chemical fertilizers systemically decrease organic matter because they activate the microorganisms that consume soil organics. High yield seeds are planted densely, which quickly consumes both the nutrients and the organic matter in the topsoil.

Building organic matter and humus levels requires managing the living organisms in the soil, similar to wildlife management. Soil management entails working to maintain favorable conditions of moisture, temperature, nutrient status, pH and aeration. It also involves providing a steady food source.[28] All the soil organisms except algae (that use sunlight for photosynthesis) depend on organic matter as their food source. To maintain their populations, organic matter must be renewed from plants growing in the soil, animal manure or other organic materials imported from off site. Adding organic material such as crop residues, manure or compost builds soil fertility as it provides humus.

Earthworms make a significant contribution to soil fertility as their burrows enhance water infiltration and soil aeration. Earthworm tunneling can increase the rate of water entry into the ground 4 to 10 times higher than fields that lack worm tunnels. Earthworms reduce water runoff, recharge groundwater and help in water retention. Vertical earthworm burrows pipe air deeper into the soil, stimulating microbial nutrient cycling at those deeper levels. Worms eat dead plant material left on top of the soil and redistribute the organic matter and nutrients throughout the topsoil layer. Nutrient-rich organic compounds line the tunnels that may remain in place for years, if not disturbed. During droughts these tunnels allow for deep plant root penetration into subsoil regions of higher moisture content. Earthworms can replace considerable energy-expensive mechanical tillage. A good population of earthworms can process 20,000 pounds of topsoil per year, with turnover rates as high as 200 tons per acre.[29]

Soil management

Industrial farming practices systemically degrade soil nutrients, organics and soil aggregation. Modern farmers continually extract nutrients and organics while replacing only a portion of the nutrients lost, but not the organics. Chemical fertilizers push plant growth but add no new organic matter or carbon to the soil, which creates compaction and accelerates erosion. Eroded soil diminishes crop yields because eroded soil absorbs 87% less water than uneroded soil, which then washes out nutrients, organics and decreases soil depth.

Adding organic material from external sources is a mammoth job and may be cost prohibitive in many areas. An acre of soil measured to a depth of six inches weighs approximately 2 million pounds, which means that 1% organic matter in the soil weighs about 20,000 pounds (10 tons) per acre. Typically, at least 10 pounds of organic material must decompose to produce one pound of soil organic matter. Therefore, to add 1% stable organic matter per acre under favorable conditions takes 200,000 pounds (100 tons) of organic material. Collecting, loading, hauling, applying and plowing 100 tons of organic material consumes considerable labor, equipment and fuel.

Soil binding substances, such as the green slime from algae, are highly vulnerable to microbial degradation. This organic matter needs to be replenished to maintain soil aggregation. Farming practices can conserve aggregates once they are formed by minimizing factors that degrade and destroy aggregation, such as:

- Avoiding excessive tillage because it mechanically breaks soil aggregation and adds oxygen to the soil which actives soil microbes that voraciously consume and deplete organic matter.

- Cultivating the soil when it is too wet which causes compaction, or too dry, which causes compaction, dust and erosion.
- Cultivating the soil with heavy equipment moving rapidly because it causes compaction, dust and erosion.
- Using anhydrous ammonia because N activates microbes and speeds decomposition of organic matter.
- Fertilizing with excess N fertilization which activates microbes up to a threshold then becomes toxic to microbes when over-applied.
- Allowing build up of excess sodium from salty irrigation water or salt-containing fertilizers.

Unfortunately, modern farmers till their fields with huge tractors that move rapidly, apply enormous amounts of anhydrous ammonia and facilitate salt invasion from irrigation and fertilizers. These modern farm practices degrade the soil ecosystem along with the microflora.

Worms aerate the soil, providing space for beneficial microorganisms but tillage may kill 90% of the earthworms and the microbes they support.[30] Tillage reduces earthworm populations by drying the soil, burying the plant residue they feed on and exposes the soil to air, putting the worms at risk of freezing. Tillage destroys their burrows and can kill and cut up the worms. Emergence times for young worms are spring and fall, typically when cultivation occurs. Chemical fertilizers, such as anhydrous ammonia along with pesticides, herbicides and fungicides are often toxic to earthworms as well as microorganisms.

Algal soil conditioners

Algal soil conditioners enhance soil structure to optimize aggregate stability to water and mechanical disruption. The size, distribution, confirmation and stability of soil aggregates are influenced by soil porosity, aeration, water movement, compaction, ease of tillage, fertilizer use, root development and erosion potential,

Microbial polysaccharides such as cellulose and chitin are the most important natural factors in the formation and stabilization of soil aggregates. Polysaccharides are simple carbohydrate structures made up of repeating units joined together by glycosidic bonds. Chitin is one of many naturally occurring polymers and one of the most abundant natural materials. Chitin provides strong structure and contributes to strong soil crusts and biodegrades in the natural environment.

Communities of microalgae develop surface veneers on soil when there is adequate light and moisture. A variety of green algae species have been used to successfully condition irrigated farmland. Several experiments have attempted to apply algae to dry land fields by aerial application. These have not been successful because algae need regular moisture to fix N_2 and too produce polysaccharides.

Soil surfaces in hot and cold deserts and semiarid steppes are commonly consolidated by microflora with algal crusts that are composed of rich organic biomass. Nutrients from algal crusts percolate into top soils with annual rains.

India has reclaimed saline alkaline soils by creating ponds which are filled with water during the monsoon that grow indigenous algae. The algal biomass is recovered and incorporated into the soil to improve the physical properties that influence infiltration, moisture retention and drainage. Algae operate as a soil conditioner that enhances water penetration which carries soil salts below the root zone where they do no damage to crops.

Smartcultures mimic nature by delivering microalgae that act as symbiants for field crops in order to solve a set of critical challenges including added nutrients, organic matter and strong soil structure.

Algae and their vibrant communities enhance soils in several ways.

Porous aggregates. Possibly the highest value soil algae provide is the production of external polysaccharide sheaths. These polysaccharides bind soil particles together in a manner that creates porous aggregates. This is particularly valuable where repeated tillage, heavy tractors and irrigation have compacted soil structure.

Algae's porous aggregates aerate the soil and enable the establishment of beneficial microflora and earthworms. These actions give the soil better insulative properties as well as a broad spectrum of amino acids, minerals and vitamins that are absorbed by crops. Improved soil organics with bioavailable nutrients result in faster growing, hardier plants that provide enhanced nutrition.

Soil organics. Algae capture and store solar energy in hydrocarbons and nutrients gleaned from the air and their moist surroundings. Crops can use immediately these building materials immediately or leave them as organics in the soil for later use. Soils that have lost their soil organics must be abandoned because they neither hold water nor support the nutrients necessary for plant growth.

Winds and water accelerate erosion of the most valued soil component first – humus. Erosion takes five times more rich organic matter compared to loam because humus tends to be lighter and often flakes in the process of organic breakdown.[31] (Think of the organic breakdown of a leaf.) Wind tends to kite the organic matter and blow it into dust clouds while water simply floats it away.

Humus loss causes clear-cut rainforest land to degrade after only two growing seasons. Rainforest land is often hilly and rocky with thin topsoil which erodes easily. Degraded soil may retain some vital plant nutrients but without water retention, the nutrients do not dissolve and become superfluous because they are unavailable to the plant.

Smartculture farmers build soil organics with cover crops, nutrient-rich algae and other microbial biomass to upgrade good soil or recover depleted soils. Growing algal biomass to enhance soil organics facilitates and accelerates the transition from chemical to organic farming by regenerating soils.

Some soils become compacted from carbonate build-up. Plants are essentially starved of critical nutrients in compacted soil when water cannot penetrate through the topsoil. Smartcultures can deliver microalgae that dissolve soil carbonates and improve soil porosity by 400%, Figure 2.1.

Figure 2.1. Soil Porosity

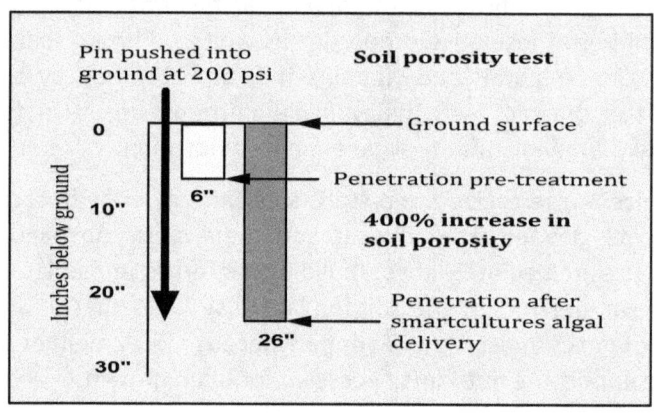

After two years of biologically conditioning soils with algae in Washington State, farmers reported production increases and lower tillage cost due to less energy consumed. Farmers used less irrigation water because conditioned soils held moisture longer. Farmers also reported that wind and water erosion was reduced substantially.

Irrigation emitter corrosion. Micro-irrigation systems suffer from a buildup of calcium carbonate that clogs lines and emitters. Emitter failure blocks water delivery and can cause damage or death to crops that do not get enough irrigation water. Algae can dissolve soil carbonate and clean the emitters.

Fungicide poisons and pollution. After each crop, farmers often must treat the field with one or more fungicides to poison nematodes and other predators. Unfortunately, these poisons also kill him the resident soil microbes needed to break down fertilizers so they are bioavailable to the plants. Fungicides not only kill valuable soil microbes but erode with wind, rain and irrigation and poison surface and well water.

Smartcultures can deliver microalgae that are benign to humans and animals but are toxic to nematodes and other soil predators. Specific algae strains are selected that rid the field of nematodes and other predators but are benign to most soil microbes. Therefore, the microbes stay in the soil to do their magic. Additional research will probably discover new benefits to field crops from algae.

Algae also play a major role as plant growth regulators.

Plant growth regulators

Plant growth regulators (PGRs) are synthetic or naturally occurring chemicals that influence plant growth, development and metabolism. PGRs are compounds that modify or control one or more specific physiological processes within a plant. Hormones are compounds produced within the plant that may be stimulated by a PGR. Plant regulators are characterized by their low rates of application because high application rates of the same compounds go by a different name; herbicides.

Algae produce antibacterial, algicidal, antifungal, antifungal and anti-protozoan compounds that are useful to crops. Blue-green algae also called cyanobacteria have produced PGR's that increased yields and vitamin C content of tomatoes, the yield and N content chili peppers and lettuce. Blue-green algae accelerated germination and growth for wheat, tomatoes, radishes and potatoes. Cyanobacteria PGR effects have been attributed to production of amino acids, vitamins and hormones.

Macroalgae or seaweed products make up nearly half of the fertilizers and growth enhancement products sold for home and small hydroponic production today. The market segments for PGR products consist of sales for use in greenhouses and intensive outdoor horticultural operations such as orchards. Seaweed products are not currently used in modern agriculture at significant scale due to their limited availability and variability in quality and price.

Discovery and development of natural PGRs that improve herbicide use or promote yields have received little attention due to proprietary considerations. In the U.S., natural processes may be discovered but not patented. Corporations are equipped and staffed to produce and test thousands of novel organic chemicals annually yet invest very little in natural PGRs because the discoveries do not convert easily to intellectual property. Unfortunately, the federal government has elected to support the chemistry of industrial agriculture research in lieu of less expensive natural processes. With negative incentives for industry research and negligible support from government, the benefits and pathways of natural PGRs remain largely unknown.

Dried seaweed meal and liquid seaweed have been evaluated for their influence on yield, quality and fertilizer use efficiency on ornamentals, vegetable, fruit, nuts and commodity crops. Positive responses included enhanced fertilizer use efficiency for treatments that included soil and foliar seaweed application. Other significant results included accelerated seed respiratory activities, increased yields of some crops, increased soluble solids in tomatoes and grapes as well as increased shelf life for peaches. Composted seaweeds added to soil increase the uptake of nutrients, probably due to a chelating mechanism in the soil. Composted seaweeds increase the bioavailability of P, N and Fe.

Crop responses to seaweeds are thought to be due primarily to cytokinins, a diverse group of PGRs that influence cell division. Some seaweeds produce antifungal compounds that prevent diseases and fungal attacks. Other seaweed PGRs have been shown to have boost herbicide impact on broadleaf weeds and to enhance the effect of synthetic herbicides by promoting herbicide uptake into the plant leaves.

Plant hormones are signal molecules produced within the plant and occur at extremely low concentrations. Hormones affect gene expression and transcription levels, cellular division and growth. Hormones regulate processes in targeted cells determine timing and formation of roots, stems, leaves, flowers ripening of fruit and the shedding of leaves. Plants, unlike animals, lack glands that produce and secrete hormones. Each plant cell is capable of producing the hormones that signal the plant shape, seeding, flowering and fruiting.

Hormones are vital to plant growth and without them plants would be a mass of undifferentiated cells – just like algae.

Not all plant cells respond to hormones but responsive cells act at specific points in the growth cycle. Plants need hormones during precise growth phases and at specific locations. Plants use different pathways to regulate internal hormone quantities and to moderate their effects. Some plants have the ability to store hormones in cells, inactivate them or cannibalize hormones by conjugating them with carbohydrates, amino acids or peptides. Plants can also break down hormones chemically, effectively destroying them, or move hormones around the plant, diluting their concentrations.

Five classes of plant hormones are made up of different chemicals that can vary in structure from one plant to the next, Table 2.1. Other plant hormones and growth regulators are not easily grouped into classes. Each class has positive as well as inhibitory functions, and most often work in tandem with each other, with varying ratios of one or more interplaying to affect growth regulations.

Table 2.1. Plant Growth Regulator Classes and Functions

Class	Function	Practical uses
Auxins	Drives shoot elongation	Increases rooting and flower formation.
Gibberellins	Stimulates cell division and elongation	Increase stalk length and flower and fruit size.
Cytokinins	Stimulates cell division	Prolong storage life of flowers and vegetables and stimulate bud initiation and root growth.
Ethylene generators	Simulates ripening	Induce uniform ripening in fruit and vegetables.

Growth inhibitors	Stops growth	Promote flower production by shortening internodes.
Retardants	Slows growth	Retards fruit tree growth.

Other plant growth regulators include brassinolides, plant steroids that are chemically similar to animal steroid hormones. Brassinolides promote cell elongation and division, cell differentiation and inhibit leaf abscission. Plants deficient in brassinolides suffer from dwarfism. Salicylic acid activates genes that produce chemicals to aid in the defense against pathogenic invaders. Jasmonates are produced from fatty acids and promote production of proteins that fight invading organisms. They are believed to play a role in seed germination, affect the storage of protein in seeds and may affect root growth.

Plant peptide hormones are small secreted peptides that engage in cell-to-cell signaling. Peptide hormones play crucial roles in plant growth and development, including defense mechanisms, the control of cell division and expansion and pollen self-incompatibility. Polyamines are basic molecules with low molecular weight that are found in all living organisms. They are essential for plant growth and development and affect mitosis and meiosis. Nitric oxides (NO_x) serve as a signal in hormonal and defense responses. Strigolactones are involved in switching shoot branching on and off. Other plant growth regulators have not been fully identified.

Modern industrial agriculture tends to substitute synthetic chemicals to spur plant growth in lieu of natural plant growth regulators. The next section examines current forms of food production.

Chapter 3. What are the Forms of Agriculture?

A field becomes exhausted by constant tillage.
 - **Ovid** (Publius Ovidius Naso), *Are Amatoria* (III, 82)

Agriculture consumes 12 billion acres of land globally, with about 70% in pasture and grazing land and 30% in crops.[32] The husbandry of feed and food plants enabled humans to diversify their diets and spend less time ensuring a food supply. For most of farming history, farmers cultivated their fields using organic production methods as they:

- Grew a rich diversity of plants (called heritage varieties today).
- Planted cover crops to replace soil organics and nutrients, hold soil moisture and minimize erosion.
- Rotated crops to improve soil fertility and minimize pests.
- Applied carbon, crop residues, animal and green manure (plant material) for fertilization to replenish soil nutrients.

Added carbon and potassium carbonate (K_2CO_3) often came from "potash," which resulted from the practice of burning wood, weeds or crop residues in a pot. The ashes were leaching with water and evaporating the solution produced a white residue called potash. Animal manure comes from draft and other farm animals that recycle organic carbon. Green manure comes from a cover crop that is plowed under to replace soil nutrients.

For roughly 11,000 years, agriculture was environmentally benign because farmers relied on natural ecological processes. Crop residues were incorporated into the soil or fed to livestock. Manure was returned to the fields where microorganisms recycled nutrients and produced soil organics. Typical mixed production farms with crops and animals were closed, stable and sustainable ecological systems that generated few external impacts. The reason traditional agriculture succeeded for multiple millennia was that this type of mixed production conserved vital nutrients.

Agriculture improved human societies, but its Achilles' heel was that food production required considerable labor, good weather, soils rich in organics and nutrients. Food production required sufficient fresh water delivered at just the right times and the avoidance of pests. When weather, soils or water failed, communities – and sometimes entire civilizations – perished. Before the last people died from community starvation, war and illness decimated the population. History will repeat starvation if regions and countries cannot manage their soils or run out of critical inputs for modern food production.

Elizabeth Kolbert, in *Field Notes from a Catastrophe,* chronicled a list of sophisticated cultures that sustained themselves for hundreds of years and then crashed when their soils no longer supplied sufficient food, including:

- Tiwanaku, Lake Titicaca in the Andes – crash: A.D. 1100, drought
- Classic Mayan civilization – crash: A.D. 800, drought
- Old Kingdom of Egypt – crash: 2200 B.C., drought
- Akkadian empire – crash: 2200 B.C., drought[33]

Jared Diamond, in *Collapse: How Societies Choose to Fail or Succeed*, describes similar eco-meltdowns that caused the Anasazi of the U.S. Southwest and the Viking colonies of Greenland to crash.[34] He shows how patterns of population growth combined with drought, over-farming and destruction of natural resources lead to deforestation, erosion and starvation. Today, we may feel insulated by our advanced technologies, but our societies are in serious jeopardy from soil degradation, overconsumption of natural resources and climate change.

Agriculture evolution

Plants growing in the wild are fertilized by natural organics (last year's plant remains, plus organics left by birds and animals). Farmers found they could increase yields of food crops by providing additional fertilizer, typically manure, wood cinders (potash) and iron slag, to promote plant growth and development.

After the Second World War, policymakers were concerned about feeding the global population of 1.6 billion, many of whom were very hungry. Around 1950, U.S. policymakers converted the munitions industry into a chemical fertilizer industry, and the Green Revolution began. Policy makers and leaders have celebrated their food production victory in the battle over nature by advances in genetics, hydrology and chemistry. The Green Revolution is likely to be viewed historically as a Pyrrhic victory because the higher yield crops came at the cost of consuming its base of nonrenewable resources. If these methods continue, many of the fossil resources farmers over-consume today will be gone, just when their children most need them.

Increased crop yields enabled a population increase of four billion people. Yet despite productivity advances, there are four times as many hungry people now than in 1950. Food producers may not be able to supply sufficient food for expanding populations. Historians may re-label the Green Revolution from green to black because modern farming is not green in a sustainability sense. Modern food production is far too consumptive of nonrenewable resources to be sustainable because it erodes its own foundation.

Classifications of agricultural methods overlap because most farmers have strong intent to grow reliable yields while maintaining good husbandry of the soil and resource inputs. The conflict occurs because modern farmers attempt to maximize their yields per unit of land through the reliance on external, inorganic inputs (compounds that contain no or minimal carbon). Government policies and incentives motivate farmers to decrease their cost of production by leveraging modern seeds, fossil fuels, fertilizers and agricultural chemicals to achieve economies of scale. Unfortunately, modern agriculture will not only exhaust fields from constant tillage but will exhaust the critical natural resources required to sustain food production.

Many modern farmers practice a variety of methods based on their fields' anticipated yields, weather and the cost of available resources. Farmers who own their land typically demonstrate better land stewardship than farmers who lease land. In the US, a majority of farmers lease their land and maximize neither soil conservation nor organic replenishment.

Industrial Agriculture

Modern industrial agriculture follows the mantra of better living through chemistry. Production methods tend to substitute synthetic products in lieu of the divine chemistry of the sub soil — microorganisms. Modern agricultural practice typically uses genetically modified (GM) seeds, extensive soil tillage with heavy equipment, massive amounts of fossil fuels, heavy application of inorganic fertilizers as well as chemical pesticides, herbicides, fungicides, bactericides and nematocides.

New hybrid and GM seeds have improved crop productivity per acre by enabling:

1. Dense plantings with negligible spacing between plants.
2. Developing varieties that were very responsive to N fertilizer.
3. Tricking the plant to put energy into fruit (seeds) rather than roots.

The new seeds were bred for productivity per acre rather than pest resistance. The GM plants are extremely vulnerable to a spectrum of vectors including competing weeds, worms, weevils, fungi, viruses, blight and other pests. Modern seeds multiply the cost of seeds because farmers cannot save their seeds under the GM contract and must pay the going price for new seeds every year. GM seeds escalate many other farm expenses because highly productive seeds consume significantly more water, require substantially more cultivation and additional protection from weeds and pests by agricultural chemicals.

Modern agriculture is a costly business, and fossil resources are the currency used to grow food. After cars, food production consumes more fossil fuel than any other sector of the economy – about 20%. Industrial agriculture depends on fossil fuels for farm machinery, food processing, packaging, transportation, fertilizers, herbicides and pesticides; see Figure 3.1. Farmers may cross a field 6 to 9 times on tractors, trucks or harvesters to produce each crop.[35] Huge, heavy tractors pulling plows, disks, cultivators, planters, spray equipment and harvesters consume enormous quantities of fuel as they crush and compact soils.

Figure 3.1 Fossil Energy Inputs Necessary for Modern Agriculture

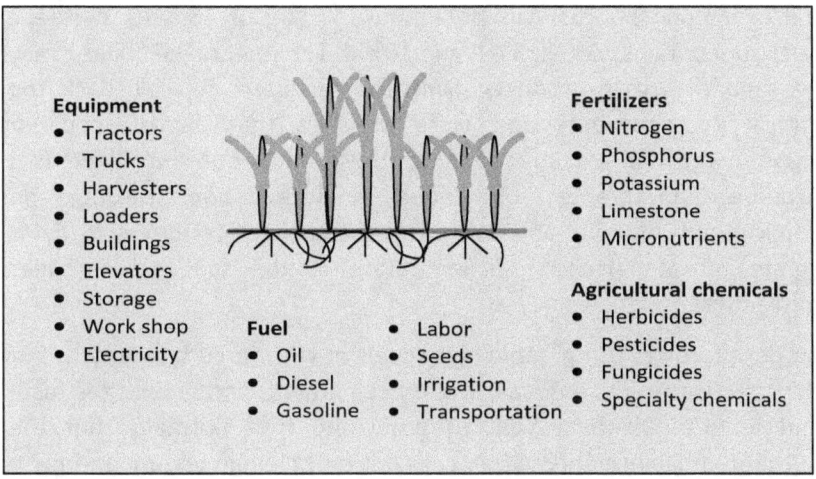

Fossil energy replaces substantial amounts of human labor. For example, assume a gallon of gasoline sells for $3 and an hour of farm labor costs $7. The hour of farm labor has a fossil fuel equivalent of about 200 hours of manpower.[36] This leverage factor has enabled substantial increases in agricultural productivity, but there is no safety net should fossil fuels not be available or become too costly.

David Pimentel and Ted Patzek analyzed the fossil energy inputs to U.S. corn production and concluded that machinery and fuel used to reduce human and animal labor totaled about 25% of the fossil energy input, while the remaining 75% is spent on agricultural chemicals to increase crop productivity.[37] Without access to fossil resources, modern agriculture could produce only a fraction of current food production because the majority of the fossil energy used goes to enhance crop productivity.

Inorganic fertilizers are mined, manufactured and applied to fields, which consumes massive amounts of fossil energy. Manufacturers use the Haber-Bosch process to produce N fertilizer that pulls N from the air using very high pressure and temperature to create anhydrous ammonia, which oxidizes into nitrates and nitrites (NO_x).

 Modern synthetic fertilizers are usually a combination of N, P, and K (nitrogen, phosphorus and potassium). A bag of 15-5-10 synthetic fertilizer contains 15% N, 5% P and 10% K. The rest, called "filler," may be sand or waste products. Synthetic fertilizers typically lack the critical micronutrients and trace minerals that plants need for vigorous growth because adding these micronutrients increases expense and creates difficulties in formulation, mixing and application. Continual addition of synthetic N-P-K fertilizers to fields creates an unbalanced nutrient state in the soil and amplifies micronutrient deficiency.

Modern farmers often apply high levels of N to force fast growth but this practice results in weak, watery cell growth and diminishes taste and texture. Plants appear to grow and fruit normally, but the nutrient imbalance and watery cells not only change texture and taste but make plants vulnerable to attack by insects and disease. The superphosphate used in synthetic fertilizers is very active and locks onto magnesium, manganese and other trace minerals, making them unavailable to plants. Chemical potassium, generally potassium chloride, is also hard on many types of plants and harsh on soils as it increases alkalinity.

The availability and cost of fertilizer marches in lockstep with the price of energy because the manufacture of 2.2 lbs of N fertilizer consumes

21.5 cubic feet of natural gas. The other fertilizer components, P and K, also require heavy investment in energy for mining, refining, mixing, packaging, transporting, storing and application. The mining and manufacture of chemical fertilizers consumes slightly over 1% of the world's total energy consumption. For many farmers, fertilizers exceed 30% of the cost of crop production.

Anhydrous accelerates the natural decomposition of organic matter which leaves the soil more compacted. Anhydrous ammonia contains approximately 82% N and is applied subsurface as a gas. Anhydrous adds acidity to the soil and require 148 pounds of lime to neutralize each 100 pounds of anhydrous ammonia or 1.8 pounds of lime for every pound of N contained in the fertilizer.[38] Anhydrous ammonia initially kills many soil microorganisms in the rhizosphere but they recover in several weeks. Bacteria recover quickly and due to plentiful N, the bacteria grow and decompose substantial amounts of organic material. When the bacteria increase and organic matter decreases, aggregation naturally declines because there is no glue produced to hold soil particles together. Compactions decreases soil porosity and oxygen which degrades the microbial communities.

Fertilizers add substantial salt to soils. Salt reduces fertility because salt ions create a plumbing problem for plants – the large ions clog the roots. Potassium chloride (KCl), also known as muriate of potash, contains 47.5% chloride.[39] Muriate of potash is made by refining potassium chloride ore, which is a mixture of potassium and sodium salts and clay from the brines of dead seas. Only table salt adds more salinity to soils than KCl. When fields become to saline, they must be abandoned because there is no practical way to clear salt from soils. In addition, many crops are especially sensitive to chloride.

Industrial agriculture breaks the natural N and P cycles and waste nearly half of applied fertilizers. In natural systems, P is recycled in a closed loop called the P cycle about 50 times in plants and animals before flowing into the sea. Modern farmers apply mined P to their fields once and most is lost to harvest, wind and water.

Supply concerns drove up P prices 800% in a recent 14-month period in 2007. Only five countries control 90% of the P mines, and China recently put a 175% tariff on their phosphate rock, eliminating exports. The U.S. already has been forced to import P, and will run out of its own P sources within 20 years. As energy prices increase and mines run out of critical agricultural chemicals, especially P, prices will continue to escalate until mined agricultural chemicals become unaffordable, probably by 2031.

Chemical fertilizers are highly soluble, which makes them ideal for plant uptake, but they rinse out of fields quickly with rain or irrigation. The chemical runoff enters fragile water ecosystems, disrupting the delicate nutrient balance. Erosion of N and P leads to entrophication, which causes an algae bloom. The algae are devoured by bacteria that consume all of the dissolved oxygen in the water. Entrophication suffocates all plants and aquatic creatures, creating a dead zone. A recent report in the journal *Science* indicated that the world now has over 400 dead zones, and they are growing in size about 10% a decade. The dead zone at the mouth of the Mississippi River is larger than the state of New Jersey.

American farmers applied over 13 million tons of N fertilizer to their fields in 2008.[40] It may be difficult to think of N as a limited fossil resource when N_2 makes up 78% of the atmosphere. However, most plants absorb N through their roots and cannot fix N_2 from the air. Some farmers use green manure by rotating their crops with plants that fix N_2, but that lowers target crop productivity because the desired cash crop can be grown only every second or third year.

Modern farmers favor synthetic N fertilizer, typically supplied as anhydrous ammonia because it is light, easy to apply and ignites immediate growth. N fertilizer is a fossil-based input because manufacturing each ton of anhydrous ammonia requires 33,500 cubic feet of natural gas.[41] About 90% of the cost of N fertilizer comes from the cost of the fossil fuels used to produce it. Chemical fertilizers enable increased crop yields temporarily, but at a very high cost in dollars, energy consumption and pollution.

While inorganic fertilizers have increased production quantity, they have decreased food quality because chemical fertilizers:

- Diminish produce flavor and texture, shorten shelf life and reduce total digestible nutrients.
- Destroy valuable soil microbes and useful insects while increasing harmful pests as well as adding toxic metals and minerals to soils, which are absorbed by produce.
- Degrade soils by increasing soil salinity, alkalinity, compaction and erosion.

In addition, fertilizer production and application accounts for 30% of farm energy use in the U.S. and is sustainable only as long as cheap fossil fuels are available.[42] Any disruption in the fossil fuel supply chain, including a spike in fuel costs will have a devastating impact on our food supply.

Industrial farmers use huge equipment to cultivate their fields and harvest their crops. Row crops such as corn have considerable spacing between rows, which necessitates cultivation before and during crop growth and development. Most GM crops are extremely vulnerable to weeds and require more cultivation and traditional varieties. The large equipment used for cultivation, spraying and harvest are not very flexible to crop variability, which encourages farmers to grow monocultures and repeated plantings at the same crop.

Modern farmers tend to ignore natural biological pest controls and instead apply millions of tons of chemical-based herbicides, pesticides and fungicides on crops to control undesired weeds and pests. These agricultural chemicals are produced using large quantities of fossil fuels and chemicals. Rachel Carlson, in her book *Silent Spring*, alerted the world that the widespread use of agricultural chemicals is harmful to life and might eliminate entire species of birds.

The nonprofit Center for Biological Diversity has put the EPA on notice that it intends to sue the agency for failing to adequately evaluate and regulate nearly 400 pesticides harmful to hundreds of endangered species as well as human beings.[43] The EPA has reopened investigations into the toxicity of the herbicide atrazine used by most growers of GM corn. Prior studies have indicated that atrazine perturbs hormones in animals and human cells and may pose a possible risk of cancer. Several new studies link the exposure of agricultural chemicals in surface waters, especially atrazine, at the time a woman conceived a baby to a heightened risk that her infant will develop a serious birth defect.[44] Other studies have shown that the weed killer disrupts sexual development in amphibians and fish and converts male frogs and fish into hermaphrodites.[45]

The unintended consequence of expanded use of agricultural chemicals, besides pollution and human health problems, can be seen in predator and plant resistance data. In 1950, only about 10 species of insects were resistant to pesticides. Today, there are over 600. Similarly, the number of weeds with herbicide resistance was near zero in 1950, while today, there are more than 400.[46] Even though insecticide use has increased tenfold, crop loss from insects is double the level it was in the 1940s – about 13%.[47] Pest resistance forces farmers to continually add more chemicals which consume more fossil fuel and poisons more waterways.

Industrial agriculture exacerbates many forms of environmental degradation including soil depletion; overconsumption of water and fossil fuels paired with air, soil and water pollution. Modern agriculture concentrates production, creating social injustice by driving out small producers. Large farms collect the majority of subsidies and use economies of scale that enable them to act as predators to family farms which undermines rural communities.

Industrial farming creates environmental and public health concerns from the systemic pollution of air, soil and water. Monocultures that are common with GM seeds diminish biodiversity and puts the entire food supply in jeopardy from a single pest or disease vector. Heavy use of pesticides, herbicides and fungicides play a role in the mass die offs of amphibians, honeybees and bats.[48] Overuse of growth-promoting antibiotics in animal agriculture leads to resistant strains that may infect humans. Antibiotics are excreted in manure, used as fertilizer and absorbed by crops. [49] The process increases antibiotic resistance in humans.

Forms of Agriculture

Organic agriculture, while not perfect, moderates both the consumptive nature of food modern production and several erosion and pollution problems.

Organic Agriculture

Peter Fossel, Raoul Adamchak, the Rodale Institute and a chorus of others recommend a return to organic farming.[50] Organic farming has been practiced for centuries. In contrast to the Green Revolution's reliance on synthetic N and inorganic P and K to maximize production, organic farmers use crop rotation, green manure (cover crops), compost, biological pest control and minimal cultivation to maintain soil productivity and control pests. Organic production limits the use of synthetic fertilizers, agricultural chemicals, livestock feed additives and genetically modified seeds (GM).

Organic farmers apply animal manure or green manure (composted plant residues) rather than synthetic or mined fertilizers. Manure contains about 80% of the original plant nutrients, but most of the nitrogen may be lost from manure through ammonia volatization (vaporization).[51] Manure must be plowed into the soil to retain the N, which takes considerable energy, disrupts soil ecology and promotes erosion. Evaporating ammonia from manure creates nitric oxides, which contributes to global warming and depletes ozone protection in the upper atmosphere.

Organic farmers grow green manure crops such as legumes (clover, vetch, alfalfa or beans) in rotation with cash crops such as food grains in order to replace soil nutrients lost to the harvest, especially N. Green manure legumes fix atmospheric N_2 in their roots in a form that plants can use. Cover crops also add other nutrients and organic matter to enrich soils, they suppress weeds and pests and moderate erosion. Cover crops usually provide only part of the necessary nutrients for the cash crop and typically must be supplemented with other sources.

Organic farming methods are internationally regulated and legally enforced by many nations, based on standards set by the

International Federation of Organic Agriculture Movements, IFOAM, established in 1972. IFOAM defines organic farming as:

> *Organic agriculture is a production system that sustains the health of soils, ecosystems and people. It relies on ecological processes, biodiversity and cycles adapted to local conditions, rather than the use of inputs with adverse effects. Organic agriculture combines tradition, innovation and science to benefit the shared environment and promote fair relationships and a good quality of life for all involved.*

Organic farming avoids the use of synthetically compounded fertilizers, pesticides, growth regulators and livestock feed additives. To the maximum extent feasible, organic farming systems rely on crop rotation, crop residues, animal manure, legumes, green manures, off-farm organic waste, mechanical cultivation, mineral bearing rocks and aspects of biological pest control the main soil productivity and killed, to supply nutrients and to control insects, weeds and other pests.[52]

The United Nations Environmental Program (UNEP) published a report in 2008 noting that organic agriculture may be the only way to solve hunger in developing countries. The UNEP report challenges the myth that organic agriculture cannot increase agricultural productivity, citing a series of studies which show that organic farming practices in Africa have outperformed chemical intensive-industrial farming. UNEP analyzed 114 farming projects in 24 African countries and concluded that organic or near-organic practices resulted in yield increases of more than 100% over industrial methods, while providing benefits such as improved soil fertility, improved water retention and drought resistance.[53]

The UNEP conclusions reinforce the findings of the 2008 report of the International Assessment of Agricultural Knowledge, Science and Technology for Development (IAASTD) panel. IAASTD is an intergovernmental process, supported by over 400 experts under the sponsorship of the FAO, GEF, UNDP, UNEP, UNESCO, the World Bank and WHO (World Health Organization). The IAASTD report stated that "the way the world grows its food will have to change radically to better serve the poor and hungry if the world is to cope with growing population and climate change while avoiding social breakdown and environmental collapse."[54] The authors found that progress in agriculture has reaped very unequal benefits and has come at a high social and environmental cost. They recommend that food producers use processes like crop rotation, organic fertilizers and sustainable agricultural practices; specifically organic farming.

Another large-scale examination of yield data from 286 farms in 57 countries showed that small farms increased their crop yields by an average of 79% using environmentally sustainable technologies, including organic farming crop rotation.[55] A 21-year Swiss study found an average of 20% lower organic yields compared with conventional farms but a 50% lower expenditure on fertilizer and energy and 97% less pesticides.[56]

Several comparison studies in the U.S. reported 10 – 20% yield disadvantage for organics. A review of comparison studies found a higher percentage of dry matter (less water weight) in fresh organic produce, which compensates for lower total yields.[57] For example, a 12 year comparison study of relative yields and composition of vegetables found a 24% lower yield but a 28% higher dry matter for organic produce.[58]

An 8-year comparison of organic and conventional fruit and vegetable production conducted at U.C. Davis showed that yields from organic, low-input practices produced equal or better yields compared with conventional, fossil-fuel-based production.[59]

A long-term study by the USDA Agricultural Research Service (ARS) scientists concluded that organic farming builds up soil organic matter better than conventional no-till farming, which suggests long-term yield benefits from organic farming.[60] Similarly, an 18-year study of organic methods on nutrient-depleted soil concluded that organic methods were superior for soil fertility and yield in a cold-temperate climate, arguing that much of the benefits from organic farming are derived from imported materials but should not be regarded as "self-sustaining."[61]

The European Union currently subsidizes organic farming, because the countries believe in the social benefits which include reduced water use, water contamination by pesticides and nutrients, soil erosion, carbon emissions along with increased biodiversity. The benefits from organic methods, supported by empirical research, are summarized in Table 3.1.[62]

Table 3.1 Benefits from organic farming

Event	Impacts
Competitive yields	Several studies show organic farm yields are equal to, or greater than, yields from farms that use industrial agricultural methods. In the developing world, yields from organic farming surpass yields from conventional agriculture by ratios of around 4.0:1.6. Worldwide, across all foodstuffs, organic farming outperforms conventional agriculture by 3:1.[63]
Conserves water	Increasing soil organic matter enhances soil moisture retention, making more water available to plants per inch of rainfall. Organic matter reduces runoff, which minimizes soil erosion and the loss of soil nutrients. Cover crops and no-till farming reduce surface evaporation and increase the retention of soil moisture.[64]
Improves soil organics	Organic agriculture builds soil organic matter, which enhances water infiltration rates and moisture retention, giving plants more water per inch of rainfall. Increased organic matter in soils reduces runoff and the associated loss of topsoil and nutrients.[65]
Saves money	Organic farming practices reduce the input costs of fossil fuels, fertilizers, insecticides, herbicides, fungicides and GMO seeds.[66]

Saves fossil fuels	Organic agriculture reduces the energy required to produce a crop by 20–50%.[67] Energy savings come from reduced consumption of water, fuel and agricultural chemicals.
Mitigates global warming	Cover crops sequester roughly 1,000 lbs. of C per acre per year.[68] Added compost doubles the amount of sequestered C to approximately 2,000 lbs. of C per acre per year, the equivalent of over 7,000 lbs. of CO_2.[69] If regenerative organic farming practices are applied to all the world's 3.5 billion tillable acres, close to 40% of global CO_2 emissions could be mitigated.[70]
Enhances biodiversity	Organic systems host a greater diversity of plant species, beneficial insects, and wildlife, which improves the ecological health of bioregions.[71]
Improves resiliency to weather variations	Organic systems produce significantly better yields under drought stress and in wet years, and comparable yields in years with favorable weather conditions.[72] Drought has a major impact on food production, accounting for 60% of food emergencies, according to a report from the FAO (UN Food and Agricultural Organization).
Increases food nutrient density	Foods grown organically typically contain more nutrients than foods grown with industrial chemicals.[73]
Reduces toxic load	Eliminating the use of agricultural chemicals in fertilizers, pesticides, herbicides and fungicides improves nutrient density and avoidance of toxic compounds for people who consume the food. Reducing added agricultural chemicals also minimizes pollution and improves human and animal health.[74]

While organic farming offers numerous benefits, several disadvantages diminish wide-spread adoption.

Organic Agriculture Disadvantages

In spite of the numerous advantages associated with organic food production, less than 1% of the world's croplands are farmed organically.[75] In 2007, 4% of the European Union's farms (where farmers receive subsidies to use organic methods) were organically managed compared to 0.6% of U.S. farmland.[76] Organic farms tend to be smaller than industrial farms, and nearly all the research on organic production has been conducted on small farms.

Organic farming is labor and knowledge-intensive whereas conventional farming is capital-intensive, requiring more energy, manufactured and external inputs. Organic farmers use labor and ecological knowledge rather than chemicals to manage to control weeds and pests. Since organic farmers cannot leverage external and synthetic inputs as efficiently, they are less able to achieve economies of scale and are typically far smaller farms than industrial farms.

Cover crops provide long-term viability to fields but reduce short-term income because growing and plowing the crop into the field consumes fuel while producing no cash crop. Using compost as a fertilizer consumes considerable fuel because it is heavy and must be plowed into the ground to avoid nitrogen volatization. Application of 5 tons of compost per acre provides 100-200 pounds of N, 30-150 pounds of P and 100-200 pounds of K. These nutrients are chemically bound within organic molecules and the N is released gradually, roughly 15% in the first year, with the remainder being released in succeeding years.[77] The transition to organic farming takes several years because it takes time to build up organic matter and release the nutrients at high enough rate to support crops.

Organic fertilizers are highly variable in nutrient composition. Synchronizing nutrient availability with plant growth and development needs is the major challenge for organic farmers.[78] Farms without livestock (stockless) may find it difficult to maintain soil fertility and tend to rely on external inputs such as manure, grain legumes and green manures. For many farms, organic fertilizers simply may not be available or affordable or the transportation costs are prohibitive.

Manure creates problems because it is a relatively inefficient fertilizer compared with chemical fertilizer. Manure contains only a modest amount of N, so a heavy application is required to replace the N removed by the prior crop. On modern farms, manure animals are often raised thousands of miles from field crops – which make transporting heavy manure impractical. Even if meat and dairy animals were close to fields, there are far too few animals to supply sufficient manure for vast grain croplands.

Some organic materials such as sewage sludge or compost made from urban wastes may introduce heavy metals such as cadmium, chromium, cobalt, led and nickel to succeeding crops. Poultry manure may contain arsenic, cobalt, copper, iron, manganese, selenium and zinc and may create harm to human consumption is large quantities are added to soils.

Washington State farmers whose crops were ruined by fertilizer laced with heavy metals have filed suit against two major suppliers of agricultural fertilizer Cenex and QFC (Quincy Farm Chemicals). The suit claims the companies regularly contracted with hazardous-waste producers and brokers to ship wastes – including heavy metals such as arsenic, beryllium, chloride, lead, and mercury – in the fertilizer. Farmers purchased the adulterated product from Cenex and QFC to disperse on their fields but were not informed about the fertilizer's poisonous contaminants.[79] Several farmers had their crops fail and their land is now poisoned and useless for decades. Poultry producer Tyson Foods, among others, is being sued by the Oklahoma Attorney General to stop spreading bird waste in the Illinois River watershed due to water pollution.[80] The National Association of Plant Food Control named a panel of regulators and fertilizer executives to craft a label for fertilizers that would disclose fertilizer's "toxic tag-alongs."

Drugs create another problem with using manure as fertilizer. For 60 years, meat producers have fed antibiotics to farm animals to increase their growth and prevent infections. About 70% of all antibiotics produced in the U.S. (nearly 25 million pounds a year) are fed to cattle, pigs and poultry.[81] Feeding pharmaceuticals to animals sustains a growing demand for meat and accelerates animal growth, but expands the presence of antibiotics in the food chain. Roughly 98% of the applied drugs are excreted as urine or manure. When manure is applied to fields, food crops absorb and concentrate the antibiotics.[82]

Organic materials from animal manure and plant residues create ecological challenges, including the following.

- **Emissions to the atmosphere** – ammonia (NH_3), nitrous oxides (NO_x), methane (CH_4), hydrogen sulfide (H_2S), odorants, dust and airborne pathogens.

- **Water pollution** – N, P, and heavy metals such as arsenic, mercury and lead, organic matter (exerting oxygen demand), siltation and pathogens.

- **Soil accumulation** – P, copper (Cu), zinc (Zn), sodium (Na) and several salts.

- **Forage accumulation** – N (fat necrosis in grazing animals), possible nitrate toxicity to animals and grass tetany, (magnesium/potassium ratio imbalance).[83]

Organic farmers often experience problems with weeds, infestations from insects, worms, fungi, bacteria and nematodes that are estimated to reduce global yields by 40% annually. Biological and mechanical controls are imperfect and organic farmers must accept a yield loss that may be larger than conventional farmers.

Many organic farmers follow organic regulations to the letter and use organic fertilizer and avoid agricultural chemicals but drop the rest of the organic production actions such as crop rotation, cover crops and composting. This strategy, called near organic farming, provides fewer soil amenities than the full set of organic actions. Some farmers burn their crop residues in lieu of composting them but burning volatizes most the N to NO_x and the other nutrients into particulates that float away from the field in smoke.

Industrial farming recidivism also creates a serious problem as some farmers have migrated back to industrial practices and falsely claimed organic production. The Cornucopia Institute, a Wisconsin-based watchdog group, has filed several complaints against large-scale dairies that have promoted their milk as organic. After a four-year investigation and legal battle, the USDA suspended the organic certification of Promiseland Livestock for four years. Promiseland is one of the largest purportedly organic cattle producers in the U.S. and manages 22,000 beef and dairy cattle in Nebraska and Missouri. Investigators accused Promiseland of violating numerous organic regulations, including feeding conventional grain to cattle and reselling conventional grain as organic.[84]

While the actions taken by organic farmers are commendable, most industrial farms in the U.S. are unlikely to convert to organic production in the near future. The U.S. would need considerably more farmland than currently exists to create sufficient organic compost. It would take several years to rebuild sufficient organic matter. Organic farming consumes tremendous amounts of fossil fuels because compost and manure must be collected, loaded, stored, transported, applied to fields and then plowed into soils in order to avoid N volatization. Organic farming saves on agricultural chemicals, but the savings often come at the cost of lowered crop productivity.

A strategy to convert modern U.S. farmers to organic would mean farmers would have to give up their GM seeds and buy expensive new production equipment to haul manure or plant residue. Since 90% of large U.S. grain farmers use GM seeds, finding a compromise that applies best practices from each form of agriculture may be the most effective strategy.

Pamela Ronald and Raoul Adamchak in *Tomorrow's Table: Organic Farming, Genetics and the Future of Food*, consider whether the world can be fed without damage to ecosystems.[85] They believe that both current and future generations of GM crops will, if responsibly managed, allow much of the world's hungry to be fed from land already degraded by the plow's slice and the tractor's compressing wheel. Creating an organic label that allowed GM seeds would serve farmers, consumers and society.

Smartcultures

Smartcultures are an extension of organic agriculture where part of the fertilizer needed is grown in algal cultures near the fields. Farmers need a cheaper source of plant nutrients and smartcultures is an appropriate combination of chemical fertilizer, organic manures, crop residues, composting and bio fertilizers. Ideally, algae recover the farm waste nutrients in the algal biomass (biofertilizer) and deliver the nutrients back to the field in irrigation water or sprayer as nano-sized algal cells that are immediately bio rowing systems near fields can overload one or more nutrients and available to the crop.

Algal cultures naturally bioaccumulate the full spectrum of macro and micronutrients crops need. G deliver them to crops exactly when needed. For example, adding more calcium when the crop is fruiting may increase the size, weight and taste of the fruit.

Drip irrigation can deliver algal biofertilizers precisely to the roots, minimizing water use and waste of water and nutrients. Algae continue to grow in the soil where soil moisture is present, which adds rich organic matter and conditions the soil, making it more erosion resistant. This model may also use no or minimal till to minimize soil disruption and provide longevity to the water-efficient drip irrigation system.

A summary of the forms of agriculture – industrial, smartcultures and organic agriculture, Figure 3.2 – illustrate variations in inputs, fertilizers, regeneration of soil fertility, pest control, produce quality and performance.

Table 3.2 Forms of Agriculture

	Industrial	Smartcultures	Organic
External inputs			
Water consumption with irrigation	Very high – flood, sprinkler; little drip; low moisture retention	Very low – with drip irrigation and high soil organics	Moderate – with cover crop and high soil organics
Seed cost and type	Expensive – GM and modern hybrids	Variable – may be GM, modern hybrids or heritage	Low – Natural, hybrids and heritage
Fossil energy use	Extremely high – tillage and chemicals.	Medium – no or minimum till; min. chemicals. a	High – compost must be plowed into soil.
Agricultural chemicals	Extremely high – fertilizers and agricultural chemicals	Modest – minimal external inputs	Low – avoids chemical fertilizers and pesticides
Fertilizer			
Fertilizer type	Chemical	Organic supplemented with chemicals	Organic – manure and compost
Nutrient source	Mined chemicals	Recovered from farm waste	Farm residues and other waste
Fertilizer availability	Not immediately bioavailable – depends on soil microbes	Immediately bioavailable plus slow release	Slow release – organic breakdown
Fertilizer delivery	Tractor and cultivation	Irrigation	Tractor and cultivation

Tillage	Heavy – fertilizer typically applied with cultivation	No or minimum till	Heavy –must be plowed into soil
Fertilizer miles	Very high – from mines, often imported	Produced near the field	High – may not be sourced locally
Nitrogen source	Haber – Bosch – high energy cost	Up to 50% from N fixing algae	Manure and compost
Cost	Very high – 30% of production	Low – recovered nutrients	Medium – if compost is locally available
Nutrient sufficiency	Yes – may oversupply nutrients; toxic	Yes	Sometimes – compost nutrient variability
Fertilizer waste and pollution	Very high – extra applied to attain yields	Low – precision nutrient delivery	Low – when compost is plowed into soil
Regenerative			
Soil nutrients	No – extraction of macro and micronutrients	Yes – adds full spectrum of soil nutrients	Yes – replenishes some soil nutrients (compost is highly variable)
Soil organics	No – extracts soil organics	Yes – adds soil organics	Yes – adds soil organics
Soil compaction	Very high – heavy tillage with heavy tractors	Low – minimum tillage	High – heavy tillage
Soil erosion	Very high – heavy tillage and soil compaction; run-off and dust	Low – minimum tillage and algal soil crusts	Low – cover crops

Salt build-up	High – fertilizers, chemicals and irrigation	Low – especially with drip irrigation	Medium – manure may add salt
Acidification **Alkalization**	High – fertilizers, chemicals and irrigation	Low – algae can moderate soil pH	Medium – compost may add acidity
Crop rotation	Sometimes yes but often no	Yes	Yes
Soil structure	Constant destruction – tillage and compaction	Builds soil structure with algae and microflora	Builds soil structure with compost
Retains soil moisture	Often no; low on soil organics	Yes – high soil organics	Yes – high soil organics
Pest control			
Natural pest controls	Low – use chemical poisons	High – stimulate natural plant hormones	Medium – use primarily natural methods
Natural pest predators	Kill them with poisons	Attract them with algal microflora communities	Attract them with compost and manure
Pesticides	High	Low	Low
Herbicides	High	Low	Low
Fungicides	High	Low	Low
Produce quality			
Nutrient density	Low – erodes nutrient density	High	High
Taste and texture	Poor – dearth of micronutrients	Excellent – precision nutrient delivery	Good – variable nutrient delivery

	Performance		
Productivity	High in short term; crashes in medium term	High	Variable. 10-20% lower yield
Sustainability	Low – will run out of fossil natural resources	High – minimal dependent on external inputs	Medium – fresh water and fossil fuels
Stability	High as long as external inputs are affordable and available	High – precision water and nutrient delivery	High – assumes availability of organic fertilizer and absence of pests
Equitability	Poor – large, wealthy will benefit; cannot afford inputs	Good – large medium small farmers benefit by increasing production and saving costs	Good – may not be practical for large farms; good for small farms
Adoptability	Poor – inputs are unaffordable for many small farmers	High – methods adoptable by all farmers	Medium –may not be adoptable by large farmers
Subsidies needed	Very high – water, power, fuel, fertilizers and crops	Medium – water	High – water, power and fuel

Several other forms of agriculture are practiced.

Sustainable Agriculture

The National Sustainable Agriculture Coalition (NSAC), a national alliance of family farm, food, conservation, rural and urban organizations promotes sustainable agriculture. The NSAC definition is: "A safe, nutritious, ample, and affordable food supply that is produced by a legion of family farmers who make a decent living pursuing their trade, while protecting and improving the environment and contributing to the strength and stability of their communities."[86] The Brundland Report states that sustainability "meets the needs and aspirations of the present without compromising the ability of future generations to meet their own needs."[87]

Sustainable agriculture integrates three main goals: environmental stewardship, farm profitability and prosperous farming communities. Sustainable agriculture came about in the 1960s as farmers and environmentalists became concerned about the impacts of agricultural chemicals on the natural environment and human health. Leo Horrigan and his team at Johns Hopkins in the Center for a Livable Future extended the definition to include more equitable distribution of high-protein food, without animal fat.[88]

Sustainable production practices involve a variety of approaches. Specific strategies must take into account topography, soil characteristics, climate, pests, local availability of inputs and the individual grower's goals. Despite the site-specific and individual nature of sustainable agriculture, several general principles can be applied to help growers select appropriate management practices, such as:

- Selection of species and varieties suited to the site and to conditions on the farm.
- Diversification of crops (including livestock) and cultural practices to enhance the biological and economic stability of the farm.
- Management of the soil to enhance and protect soil quality.
- Efficient and humane use of inputs.
- Consideration of farmers' goals and lifestyle choices.[89]

Sustainable agriculture encompasses several sets of basically organic practices, with the particular actions employed in any given situation depending on the person or organization involved. Sustainable farms produce foods without excessive use of pesticides and other hazardous chemical inputs. They often rely on alternative pest control methods such as habitat manipulation, biological controls and use of pest-resistant plant varieties. Sustainable farmers recognize the importance of protecting the natural environment and managing their farms in a responsible manner, maintaining the fertility of the land and preserving resources for future generations.

John Ikerd in *Crisis and Opportunity: Sustainability in American Agriculture* eloquently describes stainable agriculture as a movement by small farmers that may claim a label of organic, low-input, holistic, practical or real farmer.[90] Sustainable farmers try to build farming systems that are ecologically sound, economically viable and socially responsible. Sparse empirical research exists on sustainable farming independent of the research on the benefits of organic production.

Conservation Agriculture

Several international research and development organizations advocate conservation agriculture (CA) and claim the practices are a panacea for regions with poor agricultural productivity and soil degradation such as sub-Saharan Africa. CA replaces tillage with a heavy dependence on fertilizers, herbicides and pesticides. Proponents claim CA increases yields, reduces labor, improves soil fertility and reduces erosion. Unfortunately, empirical evidence shows mixed results, and it is not clear what principles of CA contribute to which effects. Concerns about CA include the decreased yields often observed, increased labor requirements when herbicides are not used, a gender shift of the labor burden to women, lack of mulch due to poor crop productivity and the common practice of feeding crop residues to livestock.[91]

Despite the publicity claiming widespread adoption of CA, the available evidence suggests virtually no adoption of CA in most sub-Saharan African countries, with only small groups of CA farmers in South Africa, Ghana and Zambia. Critical constraints to CA adoption appear to be competing uses for crop residues, increased labor demand for weeding and lack of access to external inputs.[92]

Permaculture

Permaculture is a whole systems design approach to permanent agriculture and permanent culture. Practitioners design human settlements and agricultural systems that align with the relationships found in natural ecosystems with the intent to be self-sufficient. Self-sufficiency reduces the reliance on industrial production and distribution systems that are systematically destroying Earth's ecosystems. Unlike industrial agriculture, permaculture farmers use minimal external inputs and grow diverse crops. Small-scale food production, local markets and minimal food miles provides a permaculture model.

The permaculture model evolved from sustainable food production to a holistic approach that applies to all aspects of life. The core values for practitioners include:

- **Earthcare** – the Earth is the source of all life, is our valuable home and that we are a part of Earth, not apart from it.
- **Peoplecare** – supporting and helping each other to change to ways of living that do not harm ourselves or the planet and to develop healthy societies.
- **Fairshare** (or placing limits on consumption) - ensuring that Earth's limited resources are used in ways that are equitable and wise.[93]

Permaculture has become a philosophy and a lifestyle that attracts like-minded people globally to form permaculture groups locally.

Hydroponics

Hydroponic farmers grow plants using mineral solutions and water rather than soil. In natural conditions, soil acts as a mineral nutrient reservoir, but the soil itself is not essential for plant growth. Terrestrial plants may be grown with their roots in an inert medium such as perlite, gravel, mineral wool or nutrient solution. Hoagland and Arnon found that hydroponic crop yields were no better than crop yields with good quality soils because crop yields were ultimately limited by factors other than mineral nutrients, especially light.[94] Later research showed that hydroponics offers other advantages, including constant access to oxygen, and that the plants have access to as much or as little water and nutrients as they need. Hydroponics has produced vegetables on volcanic islands that lack fertile soil.

Aeroponics, developed largely by NASA for space travel, grows plants in an air or mist environment without soil or and aggregate medium. Aeroponics culture differs from both hydroponics and in-vitro (plant tissue culture) growing. Unlike hydroponics that uses water as a growing medium and essential mineral to sustain plant growth, aeroponics cultures grow without an aggregate medium.[95] Growth nutrients are transmitted by water so, aeroponics is often considered a form of hydroponics.

Aquaculture and Aquaponics

Aquaculture farmers grow fish and shellfish that feed on aquatic plants such as algae. The Chinese have practiced aquaculture since 2500 BC. Today, half the world's commercial fish and shellfish production comes from aquaculture. Algaculture is the production of algae, kelp or other seaweed for commercial purposes. One-third of the algae grown commercially currently go to feed fish and shellfish.

Aquaponics integrates fish and plant farming. Polycultures of fish and plants may grow algae to feed the emerging fry and tiny fish. The fish urea puts rich nutrients in the water that flows into hydroponic greenhouses where vegetables and fruits grow and clean the water.

A form of agriculture that has been practiced for centuries in Asia grows algae and other microflora using primarily abundant resources.

Abundant Agriculture

Abundance: Our Future Fossil-free, Clean and Sustainable Food and Energy, (Edwards, 2011) lays out a production path with non-fossil resources.[96] Abundant agriculture makes use of Earth's oldest natural growing system – algaculture – and grows macro or microalgae that can be made into food, feed, nutrients, biofuels, pollution solutions, fertilizers, fine medicines, pharmaceuticals and vaccines.[97] Abundant agriculture follows basic organic production methods to grow water-based plants such as algae using primarily abundant inputs that will not run out – sunshine, waste CO_2 and brine water, waste water or ocean water.[98]

Forms of Agriculture

Nature makes sunshine free but not nutrients. Fortunately, algae have the capacity to recover energy and nutrients from the waste streams of farms, municipal waste facilities as well as power, cement and manufacturing plants. Microfarmers cultivate algae and possibly other microorganisms as they follow one or a combination of four sustainable and affordable food and energy, (SAFE) production paths.

Figure 3.2 Abundance SAFE Production Paths

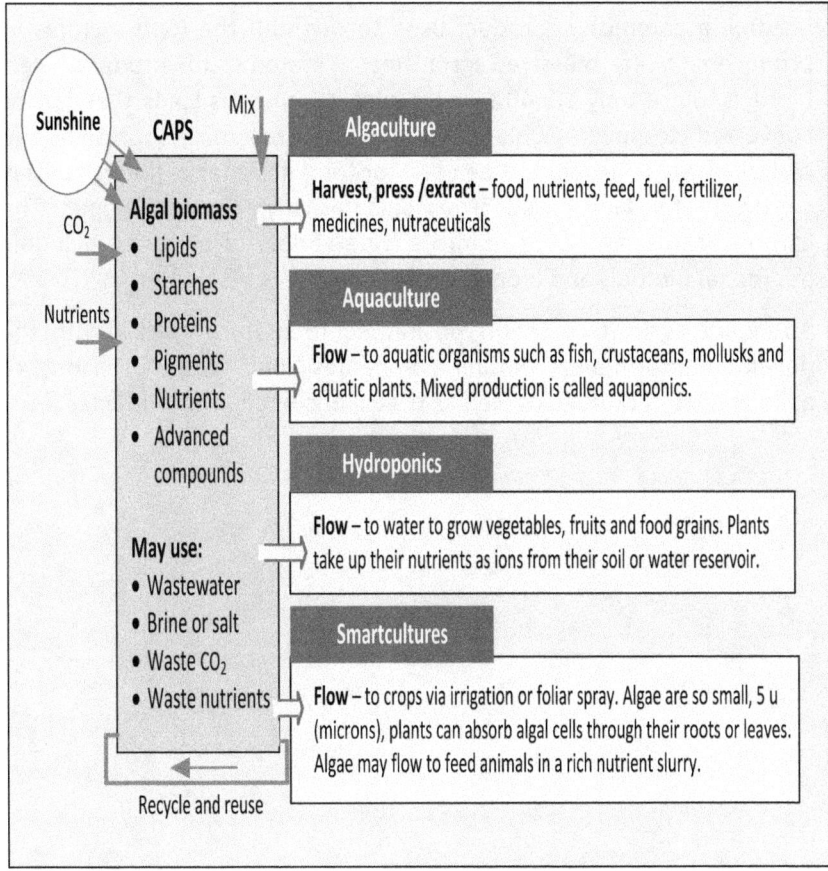

Abundant agriculture is weather-insensitive when grown in closed production systems and can be practiced anywhere on Earth. This SAFE production system overcomes many of the disadvantages of traditional fossil agriculture and offers hope for millions of hungry people.[99] Growers use 360 microfarms to produce high-quality food, feed, fertilizer coproducts using no or minimal inputs that compete with traditional crops – fertile cropland, fresh water, fuels or fossil agricultural chemicals. Production is non-pollutive and the only emission to the ecosystem is pure oxygen. While algae grow to produce food and many coproducts, the process can clean polluted water and air.

Protein and nutrients in food and feed from abundant agriculture will become a common coproduct used throughout the food systems as producers build massive agribusiness systems to produce algal biofuels. Since only 20–40% of the algal biomass is lipids that can be converted to fuels, 70% of the biomass residual is protein and carbohydrates. Protein can be used for food and makes flour that may substitute for land-based crops such as soy, corn or wheat. The carbohydrates can be made into a wide variety of products including additional biofuels and biodegradable plastic.

Abundant agriculture seems positioned to mature quickly with the development of algal biofuel companies that produce billions of gallons of algal biofuel, as well as thousands of algal coproducts.

Chapter 4. Microalgae – Nature's Nanobiofactories

There is nothing in a caterpillar that tells you it's going to be a butterfly.   **– Buckminster Fuller**

Microalgae are nano-sized plants that live in communities and capture and store solar energy in green plant bonds. Each day, algae create 70% of the world's oxygen, more than all the forests and fields combined. As algae grow, they consume two pounds of CO_2 with each pound of biomass increase. Algae synthesize roughly 0.8×10^{11} tons of organic matter daily, constituting about 40% of the total organic matter grown daily on our planet.[100]

Algae use solar energy efficiently to transform CO_2 and other nutrients to sugars, proteins, lipids, carbohydrates and other valuable organic compounds. Algae play an important role in the nutrient cycling of plants and enhance the soil structure. Some species are able to fix atmospheric N_2, some dissolve P locked in the soil, while others form compounds beneficial to plant growth and development.

Microalgae exist as particles of various sizes, many around 5 μ (microns) in diameter. (The period at the end of this sentence is about 500 microns.) While individual algae are not visible, algal communities appear first as a cloud and then tiny specks that are cell clusters. Microalgae are found in every environment where sunlight penetrates, including on, in and under glaciers, ice caps, deserts, prairies, grasslands, croplands and forests.

Smartcultures

Away from the oceans, most algal species live not in waterways but in the upper layers of soil or on rocks, grasses or trees. Algae live in symbiosis with land plants and attract a multitude of plant-useful organisms, including fungi, bacteria, yeasts, viruses, earthworms and other microfauna that build soil organics and structure in the rhizosphere.

In natural settings, algae form the critical foundation of both terrestrial and aquatic ecosystems since they make up the bottom of the food web. Algae are eaten by 100 times more creatures than any other food on Earth because they provide superior nutrition for both plants and animals. Consequently, in most ecosystems, nearly everything around algae acts as a predator.

Algae developed a brilliant strategy to survive in the presence of so many predators – grow faster than the predators can eat. Algae are the fastest-growing organisms on Earth and can double their biomass before noon each day. Algae's incredible growth rate made it the first free lunch.

Algae flourish and grow quickly in the presence of soil moisture, sunshine, CO_2 and nutrients. Cultures go dormant and rest when the soil moisture recedes or nutrients become limited. Algae attract communities of complementary soil microbes and add significant amounts of N to the soil as well as high-nutrient organics.

Algae created an ingenious strategy to survive the brutal conditions of early Earth – lightning-quick adaption. Algal cultures adapt so fast that they can survive temperature spikes, electrical storms, floods, droughts and predator invasions. Cultures can adapt in a single morning to changes in environmental conditions such as heat, nutrient limitation, population density, light availability or culture acidity (pH).

Algae growth rates and community complexity vary considerably depending on available light, moisture, temperature and soil properties such as structure and pH. A square yard of soil may contain 10 billion colony-forming units of algae. Smaller mixed algal populations occur on arid and semiarid soils with minimal plant cover.

Larger populations are found in temperate agricultural soils, grasslands, forests and other habitats where a flux of light reaches the soil surface. Even in a city, a handful of dirt may contain over 100 different algal species.

Algae grow in the soil to the distance of light penetration, which is typically several inches.[101] Rain or irrigation water may carry the algal biomass deeper into soils, where it will decompose and provide slow-release plant nutrients. Surface algae grow polysaccharide sheaths that penetrate and loosen the soil, providing room for communities of microflora. Algae grow on and in rocks, cracks and fissures and within the fabric of the rock itself.

Some blue-green algae, along with a few other species such as those of the genus Azotobacter, have the ability to fix N_2 from the atmosphere. These tiny plants propagate quickly and excrete substances which lightly bonds soil particles into aggregates. Algae also aerate the soil, which allows room for worms and microorganisms to do their work. One species, Anabaena, can act to release soil-bound P and make it bioavailable to plants. Chlamydomonas and Anabaena have demonstrated an ability to supplement or replace the inorganic fertilizer used for food crops such as corn and tomatoes.

Algal biofertilizers

Biofertilizers are communities of microorganisms that conserve and mobilize crop nutrients in the soil. This cultivated and efficient nutrient management leads to long-term sustainability and enhanced crop production by improving the nutrient and organic capital while enhancing soil structure. Algal biofertilizers constitute a perpetual source of low-cost, high-quality nutrients. Biofertilizers work in harmony with nature and form an important component of organic farming.

Biofertilizers do many things that chemical fertilizers cannot in terms of nutrients, organics and soil structure, Table 4.1.

Table 4444.1 Biofertilizers benefits over Chemical Fertilizers

Benefit	Actions
Adds to nutrient capital	• Contributes 20 to 25 kg (55 lbs) N per hector (2.47 acres) of biological fixed N_2 each year. • Amends the chemical cell properties with micronutrients. • Excretes substances that promote growth • Solubilizes (dissolve) and mobilizes phosphates locked in soil compounds. • Solubilizes and mobilizes micronutrients locked in soils. • Enhances fertilizer-utilization efficiency. • Reduces over-fertilization and pollution.
Adds to organics capital	• Amends the physical properties with organics. • Adds carbon to the soil, improving fertility, without hauling and applying compost.
Improves soil structure	• Excretes substances that bind soil aggregates such as green slime. • Increases the soil aggregate size, thereby correcting the soil compaction. • Narrows the C/N ratio in soils by adding organic biomass, which improves soil fertility. • Builds soil crusts that reduce weeds and stabilize soils, which also reduce erosion. • Enhances soil moisture retention with better soil structure and increased organic material. • Modifies saline and saline-alkali soils. • Reduces erosion from wind and water. • Minimizes agriculture chemical pollution.

These benefits have been accumulating for eons. Except for the N_2 fixing component, which was discovered by the Chinese hundreds of years ago, most of these benefits eluded discovery before the advance of biotechnology, biophysics and biosciences. Farmers are keenly aware of the value of soil nutrients and organics, but may be unaware of the substantial contribution made by algal biofertilizers because these tiny algal nutrient biofactories are not visible.

Smartcultures History

Algae have been used in agriculture for thousands of years, especially blue-green algae (cyanobacteria) for N_2 fixing and seaweed used as a biofertilizer. Seaweeds are macroalgae that grow in saline environments such as estuaries and oceans. Chinese scientists in the 11th century described the medicinal properties of *Azolla* and its use to fertilize rice.

The Romans valued seaweed as an animal feed supplement that provided micronutrients and vitamins, and they used seaweed as a soil amendment for vegetable production. The classic red color of Roman military tunics came from pigments extracted from an algae-lichen crust known as urchilles. Nearly every human population that lived near a coast harvested seaweed for the macro- and micronutrients that provided valuable vitamins and minerals.

Since prehistoric times, farmers who lived near coastlines amended soil with macroalgae to improve tilth. Good tilth refers to soil that has the proper structure and nutrients to grow healthy crops. Soil in good tilth is loamy, nutrient rich and friable because optimal soil integrates a mixture of sand, clay and organic matter that prevents compaction. Algae manufacture polysaccharide sheaths that are building blocks for the formation and stabilization of soil aggregates that give soils robust structure.

Soil components include minerals, air, water, organics and living organisms in a dynamic mix. Subsurface croplands typically include roots, other organics, rocks, silt and sand. Soil surfaces often have veneers of microalgal communities that are rich in minerals. When moisture is available, the algae propagate quickly, fix N_2 and absorb other nutrients to feed their symbiant organism and create green biomass that provides additional soil organics. The symbiotic relationship flourishes as algae provide food and foundation for crops, while plants' roots, trunks, leaves, seeds and flowers give algae a protected environment where they can thrive.

Locally Grown Nutrient Delivery

Smartcultures leverage algae's natural actions in soils as a biofertilizer and plant growth regulator that stimulates the production of natural growth hormones. Smartculture farms grow algae in a pond or container near the field, and algae enriched with nutrients are delivered to crops by irrigation or sprayer.

Biofertilizers and bioinoculants are ready-to-use, live formulates of beneficial microflora. When biofertilizers are applied to soil, roots, seeds or leaves, they enhance the availability of beneficial nutrients to the plant by their metabolic activities. Algae can be absorbed by plants through their leaves because the nano-sized cells fit through the plant's pores. Algae bring a broad spectrum of micronutrients to crops that enhance growth, productivity and quality while decreasing the need to apply chemical fertilizer.

Algae provide a booster to crops similar to the inoculums commonly given to legumes. Legumes such as alfalfa, peas and beans can fix N_2 from the air. Legumes typically produce too little N for their needs naturally because insufficient levels of the correct bacteria are present in the soil. Farmers give legumes a booster shot (inoculation) from commercially prepared sources of rhizobia bacteria to promote N_2 fixation. Inoculums are typically applied directly to the seed prior to planting or by metering inoculums into the seed furrow during planting to promote the growth of nodules for N_2 fixation.

The magic of N_2 fixation occurs when the rhizobia bacteria stimulate plant root hairs and multiply in the outer root tissue. The plant forms tissue that acts as a protective enclosure around the cyanobacteria, which is also classified as blue-green algae. The plant also supplies energy to the cyanobacteria from photosynthesis while the symbiant bacteria convert N_2 gas from the atmosphere to ammonia in the root nodules.

Algal biofertilizers work equally well on the plant's roots, stems and leaves. An algal extract sprayed on leaves was found to be superior to other organic fertilizers as foliar feed for wheat plants. A single application of algae extract 28 days after seeding increased yield 140% and grain weight about 40%.[102]

Plant growth regulators (PGRs) are natural hormones used in small amounts at specific times during crop growth and development to enhance yield and quality. The micronutrients delivered by algae may stimulate the plant to produce natural growth hormones or natural protective hormones that repel pests. Algae can carry selected hormones to the field that may, for example, boost seed germination or accelerate and extend root growth. For example, adding calcium when fruits are emerging increases the size and quality of fruits. Some soils are deficient in critical micronutrients such as iron, copper or zinc (which is critical for setting seeds) and specific nutrients and minerals can be delivered just when the plant most needs them.

Soil Conditioning

Algae transported to crops continue to grow while the field holds soil moisture. Research on rice fields shows that algae used in alkaline habitats reduce salinity by 30%, while improving pH, electrical conductivity and exchangeable sodium. Algae-conditioned soil significantly increases yields of wheat, rice, corn, peas, sugarcane, melons and potatoes.[103]

Algae upgrade the physical properties of soils such as porosity, light and water penetration, moisture retention, seedling emergence and gas exchange. After crop germination, algae may grow a soil crust composed of many types of beneficial microbes, creating a soil covering that diminishes light penetration and reduces the emergence of weeds. Algae strengthen soil stability and expand soils by growing polysaccharide sheaths. Soil expansion enables increased microbial action that enhances crop vitality.

Smartcultures

Soil conditioned with algae accelerates and extends root growth 10–100 times. Stronger and longer roots grow hardy plants that are more resistant to heat, pest, disease and drought stress. Algae-conditioned soil diminishes compaction and requires less tillage energy. Stronger soil structure makes fields more resistant to wind and water erosion. Smartcultures' combined actions make good soils better, and may replenish degraded soils with humus and nutrients while improving soil composition and configuration.

Farmers will need stronger, hardier plants that can survive the brutal impacts of climate change.

Chapter 5. Global Climate Change and Crops

So imagine a world six degrees warmer. It's not going to recognize geographical boundaries. It's not going to recognize anything. So agriculture regions today will be wiped out. — **Steven Chu,** U.S. Sec. of Energy

Temperature governs the extent and vitality of life on earth. Temperature controls the photosynthetic rates and zonal ranges of plants and animals, their habitat ranges, the metabolic rates of decomposers and the migration patterns of everything from microorganisms to the top predators.[104] The Intergovernmental Panel on Climate Change reported that the mean temperatures during the closing decades of the 20th century were higher than at any time during the past 13 centuries.[105] Higher temperatures are brutal to crops because they increase the frequency and quantity of water crops need while delivering less due to evaporation and transpiration.

Many scientists, including James Hansen at NASA, believe that global climate change is accelerating and may be approaching a tipping point, where climate change acquires a momentum that makes it irreversible. The scientific consensus estimates we may have only a decade to turn the situation around before this threshold is crossed.[106]

Global warming that included drought, hot, dry winds, wildfires and fierce storms was largely responsible for food price spikes in 2008 that caused food riots in 40 countries.[107] These insurrections disrupted national economies, spurred food theft and resulted in hundreds of deaths. Several countries created policies that prohibited food hording, waste and even food exports. In anticipation of more food shortages due to global warming and population increases, a parade of over 50 credible voices has called for a doubling of the world's food supply in the next 30 years, including:

- Robert Zoellick, World Bank president[108]
- Ban Ki-moon, United Nations Secretary General[109]
- John Beddington, the United Kingdom's Chief Scientific Adviser[110]
- LaMar Lemmons, Michigan State House of Representatives[111]
- Hugh Grant, Chairman, President and CEO of Monsanto[112]

These world leaders may be unaware that agriculture may have already peaked.[113] World food production may not have the capacity to expand at all or even to sustain current production with conventional agriculture.[114] Economists and even agricultural economists fail to factor into their business models the cost of agriculture's self-destruction, over-consumption of fossil resources and climate shocks. Nicholas Stern, former chief economist at the World Bank, examined the cost of failing to incorporate the climate change costs and burning fossil fuels and concluded the cost would be in the trillions of dollars.[115] The convergence of many factors may make fossil intensive food production unsustainable, even with expected improvements in agriculture technologies.

Modern crops are not built for climate change, especially heat, drought, dry winds and salt. Food crops were hybridized over the past 11,000 thousand years to grow within a relatively narrow temperature range, absent shocks from climate chaos. Average temperatures in food growing regions have increased $1°$ F since 1970 and 24 of the warmest years on record have occurred since 1980.[116]

Unusual summer weather in the U.S. in 2009 serves as a case study. Heat, wind and rain combined to cause severe tomato blight throughout the Northeast and Mid-Atlantic States, which threatened the entire tomato crop.[117] The "late blight," that could jump to potatoes, caused white, powdery spores, large olive green or brown spots on leaves and fruit and open lesions on the stems. Each lesion produced hundreds of thousands of infectious spores that spread on the wind, creating an explosive infection to other fields. Fungicides may protect unaffected plants from disease, but there is no cure for late blight. Many farmers had to plough their fields under, losing their entire crop. The late blight is similar to the potato blight that caused the Irish famine in 1845 that caused starvation of over one million people.

Too much heat devastates food grain yields. As temperature rises, the rate of photosynthesis increases to about 68° F and then plateaus up to 95° F. Photosynthetic activity declines above 95° F and stops at 104° F.[118] Rice, wheat and corn cannot pollinate above 104° F, which leads to crop failure. Combinations of heat, dry winds and insufficient soil moisture create partial pollination problems well below 104° F. Since pollination occurs only after the plant is fully grown, farmers must invest all their resources for an entire growing season before a heat spike ruins their crop.

Heat stress causes plants to curl their leaves to get less solar exposure and the stomata to close on the underside of the leaves in order to decrease moisture loss. Both actions disrupt photosynthesis and send the plant into thermal shock. Heat accelerates soil moisture loss which degrades crop strength, quality and yield.

David Lobel and Gregory Asner analyzed 16 years of corn and soybean data across 618 countries and concluded that each 1.8° F rise in temperature, over normal, caused a yield decline of 17%.[119] Some crops, such as rice, cannot pollinate if night temperatures stay above 80° F. Heat stress causes yield drag for most crops of 15–25%, but severe heat stress from only a few extra degrees may destroy the entire crop. Heat stress breaks crop vitality, saps energy and resistance and leaves the plants more vulnerable to disease, pests and fungi.

The heat wave in Europe in 2003 was only 2° C (3.6° F) above the long-term climatology averages but led to the deaths of an estimated 52,000 people.[120] Italy experienced a record drop in maize (corn) yields of 36% from a year earlier, whereas in France maize and fodder production fell by 30%, fruit harvests declined by 25%, and wheat harvests declined by 21%.[121] Temperature spikes, both high and low, can be extremely destructive, especially at critical phases such as germination, early growth, pollination and fruiting. High winds associated with fierce storms can destroy a crop or knock stalks around so they are too twisted for mechanical harvesters.

Higher mean or peak temperatures are only part of the story as higher lows can be equally damaging. Some food crops such as pit fruits – peaches, plums and apricots – require a certain number of cold hours, typically 450 hours below 44° F. If the weather stays warm over the winter and receives too few cold hours, the tree buds out in the spring but the buds fall off and the tree fails to set fruit. A recent study reported that chill hours have decreased 30% in central California, which has forced fruit orchards north.[122] A fruit orchard takes five years to grow trees that produce fruit, so tree farmers cannot just change to a different crop the following year.

Nobel-prize-winning physicist Steven Chu, U.S. Energy Secretary said that California's farms and vineyards could vanish by the end of the century and its major cities could be in jeopardy if Americans do not act to slow the advance of global warming.[123] He also predicted that 90% of the Sierra Nevada snow pack, on which California cities and agriculture depend, would be gone by the end of the century.

Climate scenarios for 2020 project that Mexico will lose more than one million acres of maize production to hotter temperatures.[124] Corn production in the U.S. will be forced to shift north. Temperatures in Vermont may be too warm to produce maple syrup, depriving farmers of their livelihoods.

David Battisti and Rosamond Naylor analyzed data from 23 global climate models contributing to the Intergovernmental Panel on Climate Change's 2007 scientific synthesis. They showed, with a greater than 90% probability, that growing season temperatures in the tropics and subtropics by the end of the 21st century will exceed the most extreme seasonal temperatures recorded from 1900 to 2006. In temperate regions, the hottest seasons on record will represent the future norm in many locations. More than 3 billion people live in the tropics and subtropics, many of whom live on under $2 per day and depend primarily on agriculture for their livelihoods.[125] Their analysis portends catastrophic loss of food supplies.[126] The tropics and subtropics are likely to experience the worst food loss.

Additional heat will compound food insecurity caused by variable rainfall and will increase the incidence of agricultural droughts caused by accelerated evaporation from soils, transpiration from plants, low soil moisture and high rates of water runoff from hard pan soils when it rains. Excess heat causes virga – rain that evaporates before it hits the ground. Virga frustrates farmers when fields are dry.

Australia experienced its worst drought for more than a century in 2007 and saw its wheat crop shrink by 60%. Four years later, record floods devastated over a million square kilometers in Queensland, Australia's largest coal exporter. Unprecedented 2011 January rain closed mines and destroyed crops.

China's grain harvest has fallen by 10% over the past seven years. India will soon overtake China in population and was food independent for a decade. Now, due to drought, salt invasion and failing freshwater supplies, India must import millions of tons of grain to sustain its population.

In western North America, mean annual temperatures have increased at a rate of 0.7° F a decade and 1° F a decade at higher elevations which is causing widespread tree death. The tree death rate doubled in the Pacific Northwest over a 17-year period and has doubled every 29 years in the U.S. interior.[127] Trees provide a natural CO_2 sink but when dead trees decompose, they release CO_2 into the atmosphere. Trees that are not dying are stressed and less healthy which makes them susceptible to pest invasion and wildfires. A Canadian Forest Service study found that the beetle outbreak in British Columbia has done so much damage that soon the forest will release more CO_2 than it absorbs.

Heat creates additional drag on food supplies because higher temperatures mean more crops spoil in the field and, after harvest, before they arrive at food processors. Classic, cheap food grain storage in elevators and warehouses becomes less effective due to spoilage as well as higher rates of pest infestation.

Climate change creates more extreme weather events such as the 2010 deep freeze in Florida which caused losses to farmers of $500 million.[128] Extreme winter weather in Mongolia in 2010, with temperatures plunging to -50°C, killed 1.7 million head of livestock and risks making 21,000 herder families food insecure.[129]

More severe storms will cause considerable damage to low-lying coasts and river deltas. A consensus of top climate scientists predicts stronger hurricanes (cyclones) by 2-11% with 20% more precipitation and slightly fewer occurrences.[130] An 11% increase in wind speed translates to roughly a 60% increase in damage and more rain will cause more run-off and erosion.

Global climate change deals a deadly combination punch to ocean life because much of the atmospheric carbon load dissolves in the ocean increasing acidity, while the oceans are getting warmer and oxygen levels are dropping. Current ocean acidification is moving stronger and faster than anything geologists can find in the fossil record over the past 65 million years. Andy Ridgwell and Daniela Schmidt used deep ocean core samples and computer simulations to estimate the speed and strength of current ocean acidification. Their results show acidification is taking place at ten times the rate that preceded the mass extinction of animals 55 million years ago.[131]

Highly acidic seawater becomes so corrosive it dissolves the calcium carbonate that provides structure and shelter for shellfish, corals and many other marine organisms. Small shell-building organisms provide food for invertebrates such as mollusks and small fish, which in turn become food for larger predators. Acidification dissolves coral reefs that create shelter for a quarter of the ocean's biodiversity.

Climate change most likely caused the mass extinctions in the past.[132] Numerous species are threatened as climate change disrupts food sources and entire ecosystems.

Warmer ocean water, combined with lower levels of dissolved oxygen, will diminish the entire marine food chain. The Arctic and Antarctic food chain, including whales, walruses, seals, penguins and many other birds, depend largely on krill. These tiny crustaceans (miniature shrimp) provide a high protein food source but have declined over 80% due to the warming water.[133] Krill cannot survive without the sea ice where their young hide from predators while feeding on the algae that live under the ice. Since 1974, the number of Adelie penguins in Antarctica has declined by 70%. Ten of the world's 17 penguin species are listed as endangered or threatened.

Table 5.1 Climate Change Impacts on the Food Supply

Event	Impacts
Heat	Increased temperatures causes heat stress in food crops which can significantly diminish their productivity and lead to plant death and crop failure.
Hot winds	Increased temperatures and dry winds evaporate soil moisture and increase the need for freshwater irrigation.
Water scarcity	Water, the critical resource for sustainable food production, has passed its tipping point as global warming causes food crops to need more water but water sources in many growing areas have been degraded, depleted or diverted.
Rising sea levels	Oceans will consume millions of acres of prime cropland on coasts and river deltas and tidal and storm surges will destroy millions of acres of cropland from sea salt invasion.
Ocean acidity	Dissolved CO_2 in the oceans diminish fisheries, destroy shellfish and dissolve coral reefs that protect coasts and estuaries.
Higher ocean surface temperatures	Heat provides the energy that intensifies storms, hurricanes and typhoons. Heat also changes the rainfall patterns and leads to drought and severe forest fires experienced in the western U.S.

Extended spring and fall	Spring is starting a week earlier and fall lasts an extra week, enabling pest vectors – bugs, fungi, molds, mildews, viruses and weeds – to multiply earlier and sometimes survive the winter.
Rain patterns	Shifts in rain patterns will cause huge losses of cropland that lack the infrastructure for irrigation.
Wildfires	Range lands and forests are especially vulnerable to heat, drought and winds that drive catastrophic wildfires such as those in California in 2008 and Victoria, Australia in 2009.
Loss of snow pack and glaciers	Snow packs are down 50% which means faster run off and heavy flooding in the spring. Reservoirs, creeks and rivers may be only half full when irrigation is needed later in the growing season. Melting snow packs and glaciers mean less river water for irrigation and human use.
Blowing dust	While the U.S. Midwest experienced severe flooding in 2008, Texas and Oklahoma lost millions of acres of crops to drought and blowing dust. Dust decimates crops, amplifies drought by removing soil moisture and erodes thin topsoil.
Amplified erosion	Excess heat combined with lack of soil moisture compacts soils and increases the need for irrigation and increases wind and water erosion.
Infrastructure degradation	Extended hours and more severe sun accelerates the degradation of farm infrastructure including soils, lined ditches, structures and equipment.

Human mortality	Hotter temperatures accelerate disease vectors such as HIV/AIDs and malaria and create heat stress that causes premature deaths.

Global climate change brings so many harsh variables that diminish food production that no single strategy can possibly address all the diverse impacts. In addition to climate change, several social and political factors jeopardize reliable and continuous food production.

Social climate change

Sustaining food supplies depends not only on stable climate but social stability in economics, politics, the demographics of consumption and farmers, transportation, government policies and crop diversity. Growing food crops requires considerable capital, physical labor and political policies that support food production.

Mounting debt. The vast majority of farmers must borrow capital to pay for their food production inputs. Weather variability intermittently destroys crops, leaving farmers with debt but no food. Crop failure not only increases the farmer's debt but forces subsistence farmer to have to extend their debt to buy food for the family. Most countries have no crop insurance, which means farmer's debts can mount quickly. After one or two crop failures, banks refuse to loan additional capital and the farmer most likely loses the farm.

Wars and conflict. Rising temperatures tend to spur civil wars and armed conflicts. Marshall Burke and David Lobell analyzed the incidence of African civil wars alongside local temperature and rainfall measurements from 1981 to 2002. They found a strong relationship between a spike in temperature and the likelihood of civil war. Temperature and war remained linked even after controlling for measures of wealth and democracy. Their preliminary explanation is that rising temperatures reduce crop yields and other aspects of economic activity, increasing social tension. Other studies have suggested that people become more violent when the mercury rises.[134]

New consumers. Americans have doubled their per capita meat consumption since 1990 and people around the world desire to emulate the consumption patterns of the rich.[135] The added demand for animal fodder substantially reduces available food grains. Meat consumption uses huge amounts of water since a pound of edible beef consumes 20 times more freshwater for irrigation than a pound of food grain.

By 2010, China will have twice the number of new consumers as the U.S., 600 million, who will demand higher value foods that create a high water exchange. China already leads the world in production and consumption of meat, primarily pork. Chinese per capita meat and milk consumption have doubled in the last 20 years.[136] China faces increasing demand for grains but its distressed ecology and water shortages may not support higher production. Nearly all the 1,000 lakes around Beijing have gone dry. China's two major rivers also go dry before they meet the sea. If a drought or food failure occurs, China could buy the entire U.S. grain output for a fraction of its positive U.S. trade balance.

Farmer succession. Over 40% of U.S. farmers are more than 55 years of age which jeopardizes future production unless replacements are found.[137] Farmer succession creates a very difficult problem because fewer young Americans choose to take the physical, psychological and financial risks associated with agriculture. Agricultural Land Grant Colleges are shifting from production to agribusiness because there are too few students who want to study farming.[138] Young men who decide to farm find it difficult to attract a spouse who is willing to live with the isolation associated with rural life.

Globally, other factors are driving farmers to other professions. In India, a farmer makes about $1 a day. A factory worker makes $2 a day and an educated worker may make $5–$20 a day. As more children receive education and they see other viable career opportunities, fewer choose farming.

Farmer health. HIV AIDS, malaria, malnutrition and other diseases limit many needed people from contributing to farm production. Disease vectors such as HIV AIDS tend to attack exactly the population needed for agricultural labor – young men. Failing sufficient labor, crop production diminishes or disappears.

Use of fertilizers, herbicides and pesticides has led to the death and disability of many farmers and their families, especially in developing countries. Farmers who have difficulty reading may adopt a "more is better" mantra and over apply agricultural chemicals or fail to use protective gear. The disabilities and deaths from agricultural chemical poisoning for farmers, their families and animals in several countries, including India, have been catastrophic.

Crop insurance. Farmers in the U.S. buy subsidized crop insurance to cover part of their crop, typically about 60%, in the event of crop failure or weather damage. Premiums are experience rated. In areas where crops have had damage from hurricanes, floods or wildfires, such as Louisiana and Florida, farmers cannot afford insurance. As more heat, drought and weather events occur, more farmers will be forced to leave their land.

Land redistribution. Political leaders may redistribute cropland to non-farmers who waste agricultural inputs and produce few crops. Government policies toward land redistribution, food prices and agricultural financing may undermine food production.

Food policies. Government controls on food production, distribution and pricing may make some crops impractical. Some countries, especially the U.S., have dealt a crushing blow to many farmers from subsidized food grains that are dumped on local markets as "food aid." Farmers cannot compete with subsidized U.S. grain and are forced off their farms. Canada, Mexico and several other countries have suits before the United Nations to stop the U.S. practice of dumping subsidized commodities.

Long supply chains. Fossil agriculture depends on more than 30 fossil inputs that must be affordable and available to farmers just when their crops need them. While a battle may be lost for the want of a nail, a crop may be lost for the want of a single element. The most critical fossil inputs: seeds, fresh water, diesel fuel, fertilizers, fungicides, herbicides and pesticides often have long and precarious supply chains. Long supply chains are not only expensive but risky because each fossil resource may be interrupted by supply-chain disruptions from weather, politics, war, terrorists, prices, transportation costs or other factors. A small supply disruption to fossil inputs could cause a major crop failure. Even a modest crop failure may ignite market forces to create an economic firestorm that leads to a resource run and possibly a full food cascade.

Agricultural financing. Government or private sector limitations on agricultural loans for seeds and fertilizer may prohibit many farmers from planting productive food crops. Millions of farmers are in debt because the cost of fossil inputs keeps rising. These leveraged farmers are only one crop failure away from bankruptcy.

Crop subsidies. Food imports from countries such as the U.S. that are heavily subsidized undermine local food production because farmers cannot grow food profitably at the subsidized price. American farmers have received $177 billion in subsidies over the last 12 years while several countries have become dependent on U.S. food aid and have lost their ability to produce local food crops.[139]

Subsidized U.S. crops create an unsustainable dependency. Most of Haiti's farmers went bankrupt decades ago from U.S. food aid that dumped grains on Haiti's economy at prices lower than farmers could produce. Now Haiti, traumatized by the terrible 2010 earthquake, is dependent on subsidized U.S. grain. Similarly, over 1.5 million Mexican farmers were forced to leave their land because they could not compete with subsidized U.S. corn.[140] Many of these farmers joined the flow of illegal immigrants to the U.S. from Mexico.

British International Development Secretary Douglas Alexander said "It's unacceptable that rich countries still subsidize farming at $1 billion a day, costing poor farmers in developing countries $100 billion a year in lost income."[141] In spite of world opinion, the U.S. Farm Bill passed Congress in 2008 with nearly all the subsidies intact, especially for corn and ethanol.

Subsidized water and power. Many U.S. farmers receive enormous benefit from subsidized water, water distribution and the power to pump water. Subsidizing water creates excessive water waste. Farmers in California pay only 2% of the water cost for consumers in San Francisco. When water and power are provided nearly free to farmers, they have no incentive to protect the resource. The total taxpayer cost of crop, water, water distribution and power subsidies are incredibly expensive and may be worth over twice as much as the crops produced. Unfortunately, U.S. taxpayers have no knowledge of the true cost of commodity production because they are hidden behind very, very expensive subsidies. Not only are these costs unsustainable but they mask the true social, economic and ecological costs of food production.

U.S. taxpayers would stop subsidies for irrigated corn used for ethanol if they understood that one gallon of ethanol consumed 3,000 gallons of freshwater.[142] The subsidy chain for ethanol includes the corn, irrigation water, power to pump irrigation water, 51 cents a gallon for refining ethanol a plus a litany of state and local subsidies for ethanol refineries.

Political instability. Political instability may make agricultural inputs insecure due to high prices, distribution failure or government controls on prices, tariffs and exports. Farmers lose their crops when political decisions thousands of miles away limit their access to agricultural inputs. During the food riots of 2008, raiders denuded fields before farmers could harvest their crops.

Political strife. War may conscript young men and take them out of the farm labor pool. Threat of physical harm may make working in fields too risky for farmers and cause crop loss. Some farmers cannot return to their fields because military ordinance litters their fields, especially land mines.

Monocropping. Roughly 75% of the world's food now comes from seven crops: wheat, rice, corn, potato, barley, cassava and sorghum. Farmers plant seeds from only narrow strains of each crop selected for efficiency in producing the highest yield in the minimum time and space. American's used to eat more than 15,000 varieties of apples but now less than 150 are commonly grown. The same winnowing of species applies to all food crops. Today, 99% of turkeys eaten in the

U.S. come from a single breed, the Broad-Breasted White that has been bred for meat production but can no longer naturally reproduce. More than 80% of dairy cows are Holsteins and 75% of pigs come from just three breeds.[143]

Industrial agriculture maximizes production by focusing on a few crops. Over 10 million plant and animal species have been identified but more than 80% of the world's cropland is dominated by just 10 annual cereal grains, legumes and oilseeds.[144] About 90% of the world's meat comes from only eight species of livestock. Wheat, rice and maize cover half the world's cropland and supply 80% of the total food produced worldwide.

Over 82% of U.S. farmland in 2006 was planted with the big four crops: corn, wheat, hay and soybeans. Corn subsidies motivated farmers who might have grown alternative crops to join the monocropping majority. Repeat plantings enable pests to multiply their numbers. Lack of biodiversity puts the entire production system at risk from a single vector that may be known or unknown.

A soybean disease called the Asian blight affected wide regions of U.S. soybeans in 2008 and lead to substantial crop losses.[145] Phytophthora, a plant similar to fungi, attacks hundreds of different plant species including many crops, causing tens of billions of dollars in damage per year.[146] A single vector can decimate the entire production for large growing regions which makes monocropping a risky proposition.

Global warming intensifies problems with monocropping because more heat causes:

- Higher summer temperatures that accelerate growth of nearly every vector, including weeds, weevils and worms.
- Weaker plants from heat stress that increases susceptibility to disease and predators.
- More temperate winters that enables pests to survive over the winter which gives them the opportunity for exponentially faster growth with the new crop.
- Earlier spring, giving pests an earlier start time for propagation.

In 2009, a caterpillar plague destroyed food supplies across West Africa where trillions of black, hairy larvae devoured plants, fouled wells with their feces and were so dense they drove farmers from their fields. This previously unknown species shares some similarities with army worms but has one tiny adaption; it climbs trees. Tree climbing may seem like a stupid pet trick but when farmers spray insecticides, the Liberian caterpillar stops eating the wheat, marches out of the field and climbs a tree where it is safe. The caterpillars then morph into butterflies that can fly for hundreds of miles and infect new fields. Aerial spraying is not an option because aerial application would contaminate water supplies.[147]

Monocropping creates extreme dangers and damage due to lack of biodiversity, nutrient depletion and overuse of fertilizers, pesticides and herbicides. Monocropping also accelerates systemic soil erosion adding pollution to streams, rivers, lakes and well water.

The combination of forces and factors from population expansion, global climate change, long supply chains and social, health and political impacts challenge the sustainability of food production. In spite of the need for more food voiced by world leaders, traditional agriculture may not be able to deliver even current levels of food production.

Smartculture farmers will grow crops that can better withstand climate change challenges. Recovering nutrients local to the farm will minimize supply chains, save money and cut pollution. Farmers will be able to grow crops with more biodiversity because they will have a better cost structure.

Climate change poses serious risks to crop production but the most critical threat come from the mass extinction of fossil resources. Mass extinction suggests that the end of plentiful and affordable resources will occur at nearly the same time, which will amplify the crisis.

Chapter 6. Mass Extinction of Fossil Resources

There are plenty of problems in the world, many of them interconnected. But there is no problem which compares with this central, universal problem of saving the human race from extinction.   **– John Foster Dulles**

The curse of global warming occurs from human over-consumption and pollution but a few argue it's a natural cycle. The impending mass extinction of fossil resources will occur purely from the hand of man. When the first of many fossil resources required for industrial agriculture run out locally, societies without access will face migration or extinction.

The extinction of fossil resources vital to food crops will occur sporadically because natural resources are not evenly distributed. Resource scarcity will impose a variety of impacts on farmers in different crop growing regions.

Agriculture improved human societies but its Achilles' heel was that food production was dependent on considerable labor, good weather, fertile soils, sufficient freshwater delivered at just the right time and the avoidance of pests. When weather, soils or water failed, communities and sometimes entire civilizations, perished. Before the last people died from community starvation, war and illness decimated the population. History will repeat starvation as regions and countries run out of critical inputs for modern food production.

Community starvation must be the most agonizing way for humans to die. Families are forced to watch helplessly while their weakest suffer. They see their children and elders in prolonged and excruciating misery while the human body consumes its own tissues in a desperate attempt to sustain energy. The victim's skin changes color and loses elasticity while the stomach loses its ability to digest. Eyes sink in their sockets and victims lose their memory and become weak, fatigued and disoriented. Death usually comes from diseases such as dysentery, pneumonia or heart failure that mercilessly attack the weakened body.

The mass extinction of fossil resources will occur unless we are able to make a fast course correction of our consumptive behaviors. Failing a substantial change in food production methods, fertile soils will continue to be degraded, depleted and destroyed by the joint forces of industrial agriculture, wind and water erosion, salt invasion, pest invasion and sea level rise. Fresh water will become unavailable because melting glaciers and snow packs will have caused rivers, lakes and reservoirs to go dry. Groundwater will be gone because far too many straws were drawing and wasting irrigation water. Fossil fuels will become too expensive due to peak oil and net zero exports from some oil-producing countries. Fossil agricultural chemicals will no longer be economically extractable, making fertilizers and pest control compounds unavailable or unaffordable.

Resource extinction will destroy food production in a systemic checkerboard fashion influenced by human activity and environmental factors. An extended drought in China may cause dust storms that destroy millions of acres of cropland similar to the U.S. dustbowl in the 1930s. Water wars in multiple countries will pit farmers against cities. Cities will win pyrrhic victories and the water – only to learn that farmers have no water to produce food and food prices spike. Food imports will diminish because cheap fossil fuels needed to transport food long distances will be gone. Farmers may not even have enough fossil fuels to cultivate their fields. Farmers in developed countries will find themselves in the same predicament that poor farmers find themselves in today – where they cannot afford seeds, fuel, fertilizers or pesticides.

Table 6.1 summarizes the impacts from the mass extinction of fossil resources that will occur by 2031 unless our food production systems make a major course correction and replace extraction and overconsumption with conservation and regeneration.

Table 6.1 The mass Extinction of fossil Resources

Event	Impacts
Fertile soils	One third of the world's cropland has been abandoned in the last 30 years half the remaining cropland so degraded it takes twice as much fertilizer and three times as much irrigation water to produce crops.
	• The U.S. has abandoned 245 million acres.
	• Some countries have abandoned 75% of their cropland.
	• Rising oceans will consume millions of acres on fertile river deltas and coastal plains.
Freshwater	Roughly 60% of the worlds croplands depend on extracted water for irrigation. Over 90% of water extraction exceeds natural replacement rates and many aquifers will run dry within 20 years.
	• Melting snow packs and glaciers will cause rivers to run dry.
	• More severe storms mean less water captured during runoff.
	• Hot dry winds will cause more evaporation.
	• Cities are running out of fresh water.
Fossil fuels	The green agricultural revolution has increased fossil fuel use 23 times, which is unsustainable.

Smartcultures

	• Peak oil will continue to push up fuel prices. • Zero export for fossil fuels is predicted by 2031. • Soon farmers will find fossil fuels unavailable or unaffordable. Many farmers in developing countries are unable to afford fossil fuels to produce their crops.
Fossil fertilizers	Extraction of over 200 million tons of fossil chemicals a sustainable only until economically recoverable elements become extinct. Crops accept no substitutes for fertilizers and there are no synthetic substitutes except for nitrogen. • Extracting, refining, packaging, transporting, storing and applying fertilizer consumes huge amounts of fossil energy and costs. • N is available only as long as natural gas is cheap because 90% of the cost of N fertilizer comes from the natural gas used to make it. • Economically recoverable P will be extinct within 20 years in many places. • Key micronutrients including copper, iron and zinc are predicted to become extinct by 2031.
Fossil agricultural chemicals	Several of the chemicals and poisons farmers depend will become unavailable or unaffordable. • The energy cost of production will escalate. • Society will no longer tolerate poisoning wetlands, rivers, estuaries and oceans. • Medical metrics will become more precise and identify the pathogens that lead to the degradation of human health and vitality.

92

India

India provides a useful case study in resource extinction. India will pass China as the largest population within the next decade and may have trouble supplying sufficient good food for 1.3 billion people.

The U.S. exported its "Green" Agricultural Revolution in the 1970s and sent money and technical support to India. India's government showered farmers with low-cost chemicals and new genetically modified seeds. In the years since, India has discovered that the Green Revolution is not green in the sense of supporting sustainable food crops. Today, many farmers in India can afford neither the seeds nor the chemicals to support high production crops.

India's success in expanded food production came at the high cost of increased N-P-K fertilization that increased 219, 723 and 804 times respectively.[148] Irrigation increased over 600% because transgenic crops (high-yielding, GMOs), require substantially more water because additional water improves seed germination, fertilizer absorption and plant growth. Many farmers had to drill wells and put in irrigations systems that are now dry. Even though farmers applied tons of pesticides, insects have become resistant and typically destroy larger portions of each crop than they did two decades ago.

India's government recognized that farmers needed fertilizer subsidies and began subsidies in 2005 totaling $4 billion. Rising costs for fertilizer imports and diminishing supplies increased subsidy costs to $22 billion in 2008, which has prompted calls to reform the program that India depends on to maintain its food supply.[149] India has no domestic mines for P and several other important agricultural chemicals and must pay import prices. Import prices are a function of foreign governments' willingness to export their diminishing fossil chemical reserves at reasonable prices and the political and weather stability necessary to transport chemicals long distances.

Fertilizers, pesticides, herbicides, fungicides and other agricultural chemicals have created havoc in India as monsoon rains move massive amounts of soil and embedded chemicals into waterways. Monsoon rains flood fields and enable agricultural chemicals to migrate into surface and groundwater where it poisons aquatic life, farm animals and farm families.

Over 60% of India's grain production depends on irrigated land.[150] More than half of this irrigation water comes from groundwater aquifers that are being depleted at 300 times nature's replacement rate. India's 100 million farmers have drilled 21 million irrigation wells over the last 30 years and half of those wells and millions of shallower tube wells have already gone dry. In many areas, moving water for irrigation consumes 80% of the grid energy and power outages are very common.

The Gangotri Glacier in the Himalayas has provided 70% of the water for the Ganges River for centuries but is retreating 35 yards each year.[151] When the glacier melts, the Ganges will become a seasonal river depending on annual rains depriving 40% of India's irrigated cropland and nearly half a billion people of water.[152]

Over 8 million Indian farmers quit farming during the 1990s due to rising crop input prices – seeds, fuels, fertilizers and chemicals – that created escalating farmer debt. In the decade ending in 2007, 183,000 farmers in India committed suicide because their farms could no longer provide for their families.[153] Government sources note that farmer suicides are substantially under-reported. Additional millions of farmers and family members have died or become disabled due to agricultural poisons. Trains from the city of Chotia Khurd in northern India are now called cancer trains because so many people in the farming villages must go to the city for cancer treatments.[154]

While India makes a disturbing case study, analysis in every crop growing region shows similarities, including small farmers in the US. Farmers are running out of affordable inputs to grow food. The mass extinction of fossil resources has already occurred, in a practical sense, for farmers in parts of Africa, Asia, Mideast and Central and South America because the cost of certain inputs to food production have escalated beyond the reach of all but the largest farmers.

The building blocks for sustainable food production start with fertile soil.

Fertile soil

Agriculture consumes 12 billion acres of land globally, about 50% of the Earth's land surface, with about 70% in pasture and grazing land and 30% in crops.[155] In the 60 years since the dawn of the "Green" Agricultural Revolution, industrial agriculture has produced more food but has systemically degraded and destroyed croplands. Industrial agriculture disrupts soils with cultivation, kills many of the microflora with cultivation, soil compaction and agricultural chemicals and extracts essential plant nutrients and organic matter from the soil. Loss of soil organic matter increases water runoff while diminishing moisture retention. Crops have less access to vital nutrients which causes decrease in yields. Industrial agriculture degrades the natural biota biomass and the ecological biodiversity.[156]

Soil degradation and erosion operate in a slow insidious manner that erodes farmers' ability to grow food crops. In order to maximize short-term production, industrial agriculture manipulates the natural ecosystem and disrupts and displaces precious soil, organics and nutrients. Industrial agriculture accelerates erosion around 100 times more than natural processes.[157] Soil carried away by erosion contains about three times more nutrients than are left on the remaining soil.[158] Bruce Wilkinson estimates that global erosion is occurring at a rate of about 75 billion tons a year.[159] Moving these amounts of rock and soil would fill the Grand Canyon in Arizona in about 50 years.

Over the last 40 years, over a third of the world's cropland has become unproductive and has been abandoned because it no longer supports crops.[160] Each year, another 10 million hectares (25 million acres) of cropland are abandoned due to soil degradation and erosion.[161] Half the remaining fields are so degraded they require significantly more fertilizer and fresh water to remain productive.

Erosion not only removes topsoil, organics and nutrients but carries sediment to waterways along with nutrient and pesticide pollution. Roughly 1.8 billion tons of soil is eroded from U.S. cropland each year.[162] The public and environmental health impacts from soil erosion cost U.S. citizens $45 billion annually – about 50% more than the U.S. spends yearly on health research.[163]

A single inch of fertile topsoil may take nature 500 years to create but the USDA acknowledges that over 90% of US croplands are now losing soil faster than its sustainable replacement rate.[164] Many crop producing regions have thin topsoil, only 4–6 inches deep. When the topsoil degrades or erodes, food production stops. Topsoil must be loose enough to accommodate communities of microflora and earthworms and needs sufficient porosity to enable water to percolate into the root zone.

About 40% of the world's remaining agricultural land is seriously degraded, which now require more fertilizer and water. The worst affected regions are Central America, where 75% of land is infertile, Africa, where a fifth of soil is degraded and Asia, where 11% is unsuitable for farming.[165].

Soil erosion depletes topsoil fast and furiously and without equity. Winds and water take 5:1 the rich organic matter compared to loam because humus tends to be lighter and often flakes in the process of organic breakdown.[166] (Think of the organic breakdown of a leaf.) Wind tends to kite the organic matter and blow it into dust clouds while water simply floats it away. Both actions accelerate erosion of the most valued soil component first – soil organic matter. Humus loss causes clear-cut rainforest land to degrade after only two growing seasons. Rainforest land is often hilly and rocky with thin topsoil which erodes easily. Degraded soil may retain some vital plant nutrients but without water retention, the nutrients do not dissolve and are superfluous because they are unavailable to the plant.

Rain contributes considerably to erosion. Raindrops appear to be innocent. A raindrop falling in still air arrives at about 20 mph. Storms that drive rain propel each raindrop at the speed of the wind, possibly 60 mph, at an angle. Millions of raindrops from a single storm can send soils plunging downhill and into creeks, rivers and lakes. The 2008 floods in the Midwest left washouts 30 feet deep which removed not only all the fossil topsoil but all the applied agricultural chemical pollutants, depositing them in waterways.

Raindrops smash into topsoil and break apart the soil aggregate. The individual sand, silt and clay particles change from solid to liquid and the particles fill the soil pores and reduce infiltration – water penetration. When surface pores become clogged with sand, silt or clay, overflow of water – runoff – occurs.

Water causes about 60% of cropland erosion, while wind contributes 40%. Soil erosion displays high variability based on soil type and depth, retained moisture, slope, land management such as contour plowing, crop rotation and no-till farming. Drought may accelerate the rate of erosion, especially if strong winds, rains or storms occur.

Organic farming moderates erosion because the added organic matter captures and retains more water in the rhizosphere. Water capture in organic fields may be 100% higher than in industrial fields during torrential rains. The resilience of organic fields in both extremely wet and extremely dry weather conditions creates more food security in an environment of weather extremes.[167] No-till farming also preserves soil because farmers scratch a light furrow to plant seeds rather than plowing and cultivating their fields.

Fertile croplands are also being ruined by salt invasion from irrigation and storm surges. Crops cannot tolerate salt because the large salt ions cause plumbing problems. The salt ions clog up the delicate root hairs and prevent plants from drawing soil moisture and additional nutrients. Irrigation water deposits dissolved salts in topsoil and when the water evaporates, the salt remains near the surface. Storm and tidal surges push sea water onto low-lying croplands which deposits salt and kills the crops. No economical means has been found to recover soils ruined by salt. The only practical method is to flood irrigate several times to push the salt below the root zone. However, most regions have insufficient fresh water for repeated flooding.

Expanding cities and rural development are consuming millions of acres of fertile croplands. People tend to want to live on premier croplands. Developers provide farmers a high multiple of their potential farm-based income. Arizona's croplands in the Sonoran Desert are being consumed for development at an acre an hour, 24 hours a day. The USDA expects California to lose 20% of its remaining cropland over the next 20 years.[168] Future food producers will have to find ways to improve production because by 2031, there will be substantially less available fertile cropland.

Freshwater

Food production depends on fresh water – lots of water. People consume about one gallon of water per day in their beverages but growing food for one day requires about 500 times more water, 528 gallons.[169]

Over one-third of productive cropland depends on irrigation. Irrigation delivers about 2,600 km^3 of water to the land surface each year, or about 2% of annual precipitation over land.[170] Globally, irrigation for crop production claims about 70% of all freshwater and about 80% in the US. The history of irrigation is replete with failures of cities and societies due to soil waterlogging, salt invasion and depletion of water supplies. With the demand for water growing steadily among the major consumers – agriculture, residential and industry – competition is intensifying.[171] In water wars, farmers almost always lose to money – cities and industry.[172]

Our planet held far too little fresh water to support food production for expanding populations even before global warming began melting and evaporating our ice caps, glaciers, snow packs and reservoirs. Our faucets and fountains are already going dry because farmers extracting stored fossil groundwater reserves that were laid down millions of years ago. The aquifers on which food production depends are being depleted rapidly. Many aquifers will go dry in this generation and some have already crashed.

Water creates the primary limitation to food production because failing available fresh water; crops quickly wilt, stunt or die. Without sufficient fresh water delivered on time, crops fail and the land reverts to its natural state – which in much of the world is prairie or desert – and human populations must migrate to plentiful water.

Production and yield are directly related to water use. Insufficient applied water stresses crops and decreases yield. More irrigation has doubled food production over the last 30 years but at the unsustainable expense of tripling the freshwater consumed.

Much of the 300% increase in water consumption occurred because new croplands expanded into deserts. Desert regions are productive due to the considerable solar energy but the heat consumes more water from transpiration (plant water losses) and soil moisture evaporation. Irrigation systems often lose 50% of the applied water before the water reaches the crops from leaks and evaporation.

Today, when the number of hungry people has reached record highs in America and the world, acute water scarcity has struck countries in the Middle East and North Africa, as well as Mexico, Pakistan, South Africa, the United States and large parts of China and India.[173] Iran was forced to import over a million tons of grain from the U.S. in 2007 because their crops failed due to heat and drought.

Water is incredibly expensive to move and delivery costs of $4,000 or more per acre are typically heavily subsidized by governments.[174] Surface water supplies are typically far distant from croplands and require building, operating and maintaining huge dams and reservoirs connected with hundreds of miles of canals, pipelines and pumps that deliver water. Groundwater irrigation requires pumping and as water tables drop, pumping becomes increasingly more expensive. Many farmers have found the cost of pumping groundwater to exceed the value of their crops and have had to abandon their land.

The U.S. faces severe freshwater shortages too. U.S. Geological Survey shows California's San Joaquin Valley has lost 60 million acre-feet of groundwater since 1961. About 20% of the groundwater pumped in America comes from under the Central Valley.[175] In normal years, California produces more than half of the vegetables and fruit consumed in the U.S. The state irrigates 9.6 million acres, using roughly 34 million acre-feet of water from lakes, reservoirs and rivers or pumped from groundwater. Lester Snow, director of the California Department of Water Resources, said in 2009 that California may face its worst drought in recorded history.[176] The Central Valley Authority that distributes irrigation water through the heartland of California announced a zero allocation to many crop regions. The Bureau of Reclamation estimated that one million acres would be put out of production and another two million acres would grow less food than normal. Lester Snow called the situation grim because the more than half the snowpack that usually stores California's water melted in the heat.[177]

Interior Secretary Ken Salazar announced in July 2009 that $40 million in federal stimulus funds would be spent to pull more water from depleted underground aquifers in drought-stricken California, despite clear evidence that the well-drilling will degrade the quality of water delivered to millions of residents. The stimulus money will dig 50 new wells, retrofit another 40 old ones and install temporary pipes and pumps to move water to crops and orchards.[178] While everyone agrees the state's aquifers are quickly being drawn down, still no rules govern how much water can be pumped out.

Federal and state governments also allow landowners with wells on their property to swap their underground water for higher-quality canal water. The brackish groundwater is then pumped into the aqueduct that supplies cities, farms and industry to the south with higher saline water. The saltier water will cause $18 million in economic damage such as problems with water recycling plants and the deterioration of appliances.[179] Pumping lowers the water table, drying up springs and reservoirs. Drawing water creates subsidence and sink holes which break bridges, building, roads and canals.

Drought also plagues the South and Eastern U.S. In the fall of 2008, Texas farmers received no fall rain and lost their winter wheat crop. Farmers were not able to plant spring crops because there was no soil moisture. Texas farmers rediscovered that if there is insufficient water to germinate weeds, crops don't grow either.

Rational government policy would limit irrigation to sustainable yields from surface sources and groundwater held in aquifers. Instead, government policies in the U.S. and globally have encouraged maximizing short-term food production by subsidizing water including transport, delivery and the energy needed for pumping. When a commodity has a near-zero cost, users waste it. Inefficient and over irrigation wastes trillions of gallons a freshwater each year. Over-pumping at several times the sustainable yield has resulted in plunging water tables on every food growing continent. Many aquifers are falling at 10 feet a year and several major aquifers will crash before 2031.

Fossil fuels

In 1940, farmers produced 2.3 calories of food energy for every calorie of fossil fuel energy inputs. Industrial farming substitutes fossil fuel energy for human labor and consumes 10 calories of fossil fuel energy to produce each food calorie – a 23 times increase in fossil fuels.[180] Modern farming in the U.S. consumes ten times more energy than it provides to society in food energy.[181]

Since 1981, oil extraction has exceeded new discoveries by a continually increasing margin. Global oil consumption in 2008 was 31 billion barrels of but new discoveries amounted to only 7 billion barrels. World oil reserves are now in decline, dropping every year.[182] New oil fields typically lie in deep oceans where the oil is very expensive to recover and current technologies can only recover about half as much oil compared with oil pools under continents.

The debate on peak oil misses a far more perilous concern – net zero oil exports. Net zero exports occur when exporting countries decide to use the oil themselves rather than sell it on the open market. As oil reserves shrink, oil exports are likely to decline sharply because they have more value in-country rather than on the world market. Net zero oil exports means there is no oil to buy.

The problem for fossil fuel consumers occurs because the cash infusion from oil exports in Venezuela, for example, stimulates domestic consumption of government subsidized 19 cent a gallon gasoline. As reserves fall, oil prices rise, bringing in more cash and further increasing domestic consumption, to the detriment of exports.

Geologists Jeffrey Brown and Samuel Foucher built economic models for net exports that show once oil production in an exporting country peaks and begins decline, exports drop precipitously.[183] Due to increased domestic demand, only about 10% of post-peak oil production is exported. Why should a country export crude oil when they can use the oil domestically to create products such as plastics that are worth 5 to 10 times more than crude oil? Their most likely case scenario predicts that the top five oil producers will approach net zero exports around 2031.

About 78% of the corn seeds planted in the U.S. are genetically modified to be more productive but consume substantially more water and fossil fuels.[184] Transgenic seeds cannot compete with natural seeds or pests and must be protected with extensive cultivation and expensive inorganic chemicals in fertilizers, herbicides, pesticides and fungicides. Seed production, whether genetically modified or hybridized, also consumes extensive fossil fuels. In addition to heavy fuel consumption for crop production, massive amounts of fossil fuels and resources are needed to:

- Drive the tractors and trucks to transport food to processors.
- Convert crops by food processors into forms of food desired by consumers.
- Produce and deliver food additive such as vitamins, minerals, emulsifiers, preservatives and colorings.
- Transport food to retailers who may be thousands of miles away.
- Deliver just-in-time fresh or refrigerated food in trucks.
- Produce food packages and deliver the necessary boxes, cans, paper, plastic, glass, metal and sealing compounds.

Agriculture has supplied humans with food for thousands of years. Yet the substitution of fossil fuels for human and animal labor in the last 60 years makes modern food production totally dependent on increasingly scare fossil resources and threatens to annihilate itself.

Fossil agricultural chemicals

Farmers currently use over 207 million tons of mined inorganic fertilizers annually which are sustainable only as long as all the ingredients in fertilizer are economically recoverable.[185] Fertilizer production accounts for 30% of farm energy use in the U.S. N fertilizer production uses the Haber-Bosch process, Figure 6.2, with N_2 gas and H_2 gas, over an enriched iron catalyst, to produce ammonia. Global industrial N fertilizer production costs about $15 billion a year and consumes about 2 million barrels of crude oil every day.

Figure 6.2 Nitrogen Fertilizer Production

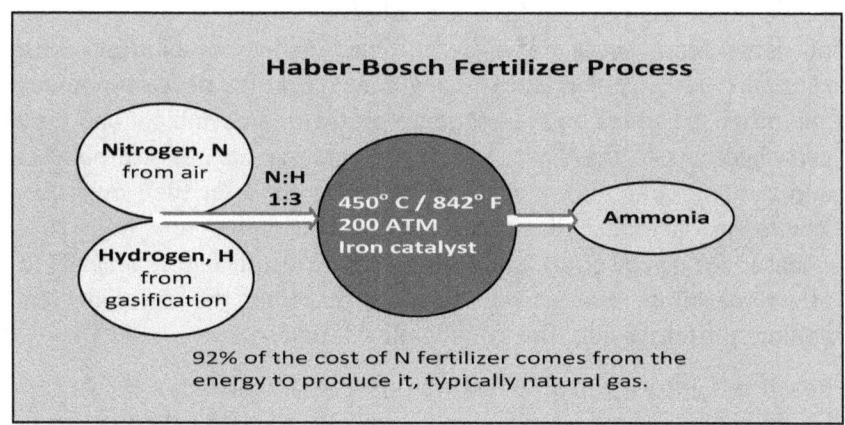

Production of reactive N, nature's most promiscuous element, has increased by a factor of 12 since 1960, to about 400 billion pounds by 2005. About 80 percent of this N is used in crop production.[186] Synthetic N production is over double that of all natural processes on land combined. In the U.S., people consume only about 10% of the N farmers apply to their fields every year.[187] The remainder ends up in the environment.

Fertilizers are not sustainable because they depend on fossil fuels for mining, transport and application. Several fertilizer ingredients will peak in a generation. Phosphorus, copper and zinc are likely to become unavailable or unaffordable, even before the extinction of fossil fuels.

Our food supply system is not sustainable because it is dependent on massive amounts of fossil fuels, industrially fixed N and mined inorganic chemicals. Our food supply has become dependent on the wisdom, benevolence and cooperation of a few heads of state and petroleum company executives.

Every organism including a single cell, plant, tree or human requires N for its physical structure, function and reproduction. Although the atmosphere is made up of 78% N, plants cannot use atmospheric N directly. The N must be converted into nitrates or ammonia, which is often accomplished by N fixing bacteria and algae. Plant nutrients converted to bioavailable form are collectively referred to as reactive nutrients. Many species of algae and bacteria live symbiotically with specific types of plants such as legumes. Legumes develop nodules and other structures on their roots to protect their symbiant and feed carbohydrates to the plant. Free living bacteria such as Azotobacter and the blue-green algae such as Anabaena fix N for their own use. When the bacteria and algae die, they decompose and their N is released for use by crops. Soils often contain insufficient reactive N to attain maximum productivity. Farmers make up the difference by adding substantial quantities of chemical fertilizers.

Reactive N enters crop production from several sources:

- Industrial production of synthetic fertilizer that combines natural gas and N_2 to produce ammonia.
- Microbe-driven decomposition of organic matter.
- Bacterial N_2 fixation, the process in which microbes, often associated with legumes such as soybeans and alfalfa, break the N_2 bond.
- Lightning that can split the N_2 bond.

Crops vary in their ability to absorb N but none take in the entire available N. The degree to which crops utilize N, called N use efficiency (NUE), is measured in crop yield per unit of added N. Farmers in the U.S. average about 150 pounds of N per acre for corn and NUE typically falls below 50% as well as for most other major crops.[188] This means more than half of all added reactive N flows from the field into the surrounding ecosystem.

The excess reactive fertilizer is mobile and rides on air and water from the field. Additional N in the form of nitrate seeps into drinking water, where it lurks as a health risk, especially for pregnant women and children.[189] Large amounts of N volatizes into the atmosphere as ammonia where it creates smog and causes respiratory disease. Agriculture is also the largest human caused domestic source of nitrous oxide, a highly reactive form of N that contributes to global warming and reduces the stratospheric ozone that protects us from ultraviolet radiation.

Fertilized soils release more than two billion tons of greenhouse gases every year, especially CO_2, methane and nitric oxide. A recent scientific assessment of nine global environmental challenges that may make the Earth unfavorable for continued human development identified N pollution as one of only three problems – along with climate change and loss of biodiversity – that have already crossed a boundary that could result in disastrous consequences if not corrected.[190] A review by the National Institutes of Health suggests that elevated nitrate concentrations in drinking water raise the risk of cancer, Alzheimer's, diabetes and heart disease and drives up mortality rates.[191] Ecological pollution in waterways enables the spread of invasive species such as ragweed that elevates pollen pollution and mosquitoes and snails that carry infective agents.[192]

The EPA announced in 2009 that a 45% reduction in N pollution was necessary to have a substantial impact on the dead zone in the Gulf of Mexico. Increased crop production using current methods will aggravate rather than moderate the poisoned ground water and dead zone problems. A 2007 Iowa Department of Natural Resources report indicated 274 Iowa waterways were seriously polluted. Fertilizer run-off causes such a problem that Iowa was forced to install the largest and most expensive nitrate removal plant in the world.[193]

Farmers put tons of fossil fuel-based herbicides, pesticides and fungicides on crops to control undesired weeds and pests. Modern genetically engineered (GE) crops are more productive but require substantially more chemical poisons for protection from pests. The annual health and environmental cost from pesticide use in the U.S. is estimated to be in excess $12 billion.[194] Agricultural chemical exposure creates a serious lag effect for human and animal health that has not been factored into health or environmental impacts. The lag effect occurs because many illnesses such as cancers and respiratory diseases from chemical exposure will appear in the future.

Fossil nutrient sink. Organic farmers rotate crops with green manure cover crops that replace some of the nutrients lost with harvest. Crop rotation also helps with pest management, builds soil organics and helps with water retention.

In contrast, industrial farmers replant the same crop, such as corn, each year because they can make more money. They apply inorganic fertilizers to approximate the soil's natural fertility. Unfortunately, both crops and pests develop resistance. Food crops require increasingly more fertilizer to sustain the same level of production. Repeat plantings enable invasive weeds and pests to propagate which creates a significant drag on crop growth. The desire to improve crop productivity combined with rising pest resistance demands more fossil intensive fertilizers, herbicides and pesticides.

Industrial farmers skip the replacement of soil organics with compost because substituting fossil fuels and mined compounds enables high productivity – in the short term. Each year a farmer begins preparing the land for a new crop by replacing last year's lost nutrients with additional chemicals from degrading mines.

The fossil nutrient problem illustrated in Figure 6.2 occurs because a farmer applies 100 pounds of P, or other fossil resource, per acre but loses about 80 nutrient pounds each year. The crop harvest removes about half and that P enters the human food or animal feed chain. Another third may be lost to erosion from wind, rain and irrigation and moves from the soil to nearby waterways where it stays in dilution or is harvested by algae or other water plants. In both cases, harvest and erosion, the nutrients are lost to the field. Each year the farmer begins again with a degraded field and uses fossil fuels to cultivate the field, inflicting more soil erosion and consuming additional fossil inorganic chemicals for fertilizers, herbicides, pesticides and fungicides.

Plants cannot be fooled by substituting other elements and there are no synthetic substitutes. After mines of P, manganese, zinc, copper and other vital elements are depleted, the only available source will be recycled waste streams or ocean water. Neither option is used because those sources are too salty for direct use on land and conventional extraction methods require far too much energy for economic recovery.

Figure 6.2 Loss of Fossil Nutrients – Resource Sink

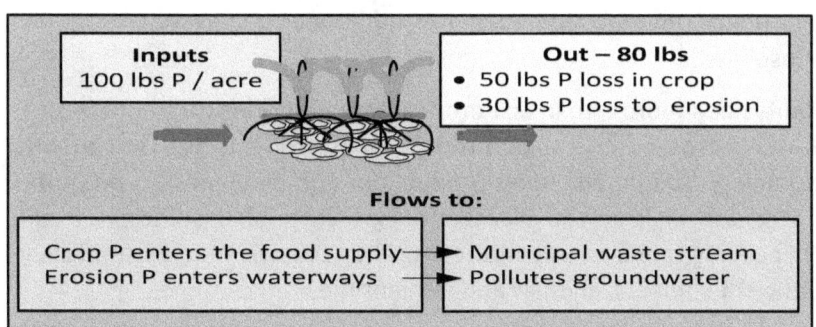

Phosphorus. P is an irreplaceable ingredient of life and, with the exception of water, is the most critical limited resource for crops because P sustains all living cells. Phospholipids form and maintain cell membranes and serve as the key structural components of DNA and RNA. P provides shape for DNA, which provides the blueprint of genetic information contained in every living cell. A sugar-phosphate backbone forms the helical structure of every DNA molecule. The element also regulates ATP (Adenosine-5'-triphosphate), which is the main energy storage and transfer molecule in cells. It is also necessary for the formation and maintenance of bones and teeth in animals and humans. The human body contains about 650 grams of phosphorus, most of it in our bones.[195] About 20% of the human skeleton and teeth are made of calcium phosphate, $Ca(H_2PO_4)_2$.

The phosphate ion combines with various atoms and molecules within living organisms to form many different compounds essential to life. Unfortunately, P is one of the most chemically reactive nutrients and readily transforms to a variety of compounds with no bioavailability to plants. P reacts quickly with air and other oxygen-containing substances and does not occur naturally in its elemental form. P exists in small concentrations in many different minerals, which make it expensive and energy intensive to mine. It occurs as a charged group of atoms or an ion made up of a P atom and four oxygen atoms (PO_4)

and carries three negative charges. Miners do not mine P; they mine phosphate minerals and then use additional energy to extract the phosphate.

Plants need more P than any other element besides carbon because P drives photosynthesis, sugar production, (stored energy), root growth, blooming, fruiting and seed production. The element also promotes N_2 fixation in legumes. Plants stop growth with insufficient P and science has found no substitutes for P in cellular metabolism. Failing sufficient P, plants, animals and humans die.

Of the total P in soils, typically only 0.1 % is plant available because most P is locked in insoluble molecules and cannot be absorbed.[196] A P deficiency severely restricts plant growth and yields. Farmers address P deficiency with the application of chemical P fertilizers but this practice is expensive, pollutive and destructive.[197] Repeated application of chemical P fertilizers leads to the loss of soil fertility by disrupting microbial activity and diversity, which reduces crop yields substantially.[198] The application of P reduces microbial respiration and metabolism which effectively stops growth.[199]

Industrial farmers applied 170 million of tons of mined P on their fields in 2008 to produce their crops. When the crops are harvested, half of the P is lost with crop and flows in the human or animal waste stream where it is typically burned, buried or released into rivers that flow to the ocean. Consumers eat foods rich in P but only about 1% of the P is retained in adult bodies and 99% is flushed down the toilet. More P is retained by children whose bodies use it for growth, energy and body structure. Much of the P remaining in fields after harvest erodes and becomes a pollutant in surface and well water.

Massive global P consumption for food production depletes this precious fossil resource rapidly. Fertilizers are made up of N-P-K and the price of each of these components directly affects the prices of the other two. Only five countries control 90% of the world's phosphate, which creates an oligopoly pricing model. Each year, P costs more to extract because mines are deeper. Many P mines are degrading as the easily mined, high quality upper layers have been mined. The more costly extraction of lower quality rock necessitates extracting more phosphate rock to produce each ton of fertilizer.

The P in phosphate rock has no bioavailability (uptake capacity) for plants, so it must be refined into a form that plants can absorb, phosphoric acid, H_3PO_4. Refining P depends on the availability of sulfuric acid, which is only produced by only a few developed countries.

Each ton of phosphoric acid consumes:
- 3 tons of sulfuric acid.
- 3.5 tons of good quality phosphate rock.
- Extensive energy for mining the sulfur and phosphate rock.
- Considerable energy for making the sulfuric acid and refining the phosphate rock.

The few developed countries that produce sulfuric acid can control the price of P by manipulating the price of sulfuric acid. Oil-producing countries can also affect the price of P by changing the supply of oil exports because the P supply chain depends on cheap fossil fuels.

Consumption of P is increasing about 3% annually and about 40% of the world's P comes from one country, Morocco. The P mines are in Western Sahara, a disputed independent territory, occupied by Morocco to exploit the P mines. China, the country with the largest P reserves, recently imposed a 175% tariff, effectively eliminating exports. The U.S. has only 12 P mines, and 75% of the domestic P comes from a Florida mine, which is declining rapidly and will run out within 20 years. The U.S. exported P for decades and now imports about 10% of its P, from Morocco. The U.S. will need to increase imports annually as domestic mines degrade and production diminishes. Europe, Scandinavia, Indonesia and India have no P mines and must import their P.

The lack of P availability or affordability will create a P famine, Pfamine. The Pfamine crisis does not start when phosphate rock reserves become extinct. Pfamine starts when the demand for P exceeds the supply of phosphate rocks available for export. Farmers living in countries that do not have domestic mines may desperately need P fertilizer but none may be available. The Pfamine will plague developing countries without P mines first because they will find P unavailable or unaffordable.

There is no artificial substitute for P in agriculture. As world reserves of this critical natural resource diminish, prices will skyrocket. Scientists at Linköping University in Sweden and Arizona State University predict peak P will occur about 2031. Peak P will pose a serious threat to agriculture as global reserves of high-quality phosphate rock go into terminal decline.[200]

The nature of fossil resources

The nature of fossil resources may be illustrated by their shared characteristics.

Table 6.2 Shared Characteristics of Fossil Resources

Characteristic	Description
Limited supply	When reserves run out, the supply is gone.
Easiest extracted first	The easiest resources, closest to the Earth's surface, are extracted first.
Last half harder to extract	The second half of a reserve may cost 5–10 times as much to extract as the first half.
Supply plummets and price skyrockets	Long before the resource runs out, high demand for the limited resource drives prices up exponentially.
Nationalism	Countries with primary sources limit access. Some countries restrict exports – decreasing supply which drives up world prices further.
Regional stock outs	Some regions, especially poor countries, are forced to go without the resource.

No substitutes	Many fossil resources have no effective natural or synthetic substitutes.
No manufacture	Fossil resources cannot be manufactured at all or, in some cases, not at a practical price.
Resource fights	Competitors fight over the last tidbits of each diminishing fossil resource – especially if it means food versus hunger for their family and community.
Fear	Consumers develop a fear mentality that the precious resource will be unavailable to them.
Hoarding	Fear causes suppliers, buyers and greedy speculators, to hoard the resource causing further shortages, more fear and escalating prices.

Access to fossil resources depends on supply, location, cost of production, transportation and a wide range of environmental, social and economic factors. Fossil reserves and years to exhaustion are countable and predictable because mining geologists have explored and mapped the planet for fossil resource reserves that are economically feasible to recover. Two serious factors that are not predictable include a supply disruption and a resource run. Both can have catastrophic consequences for food production.

A supply disruption may occur due to fierce storms that impede mining or transportation, natural disasters such as earthquakes or volcanoes, political actions such as nationalization of mines or trade embargos, terrorist actions or war. Over 40% of the world's P comes from mines in Morocco and if anything were to happen to those mines, world food supplies would plummet. Since P is also used for munitions, countries would bid against each other driving up prices of remaining supplies.

Fear of stock outs will lead to a resource run, Figure 5.3 that operates like run a bank where suppliers, farmers and speculators buy remaining stocks and hoard whatever supplies are available in anticipation of even higher prices. A resource run amplifies an already difficult supply situation and has two serious impacts. First speculators and hoarders push prices beyond levels farmers can afford. Second, it removes product from the market which fuels fear and transforms a tropical depression into a category 5 hurricane.

The disturbing detail about a resource run is that a tiny, apparently innocent action can fracture the tripwire and ignite a devastating resource run. The resource does not have to actually be in short supply; just the perception of a supply shortage that fans the flames. The tripwire may be broken by a real or perceived stock out, a supply disruption, government policy limiting resource exports, war, terrorist actions or inclement weather. Once the run starts, rumors proliferate, fear of loss drives people to do stupid things and the cycle intensifies.

Regional stock outs of critical fossil inputs will precede a total end of supply. Poorer countries and poorer farms will be the first impacted by dwindling fossil resources which means regional starvation will occur from local crop failure before mass starvation. However, a single shock from weather, politics or war could interrupt resource supplies and lead quickly to mass starvation because far too little food is stored to support large populations in mega cities.

The predicament of farmers in Bolivia provides a case study on spot outages. In 2009, Bolivian farmers were forced to watch their soy beans rot in the fields due to the lack of diesel fuel.[201] Neither farmers nor truckers could find enough diesel fuel to take the crops to food processors. Many farmers were forced out of business, not because they failed to make all the investments necessary to grow their crops, but because government policies created a shortage of diesel fuel. Bolivia, the poorest country in South America, has banned biofuels because they compete with food and create severe ecological damage. The Bolivian government may join other governments in restricting exports or nationalizing its energy and fertilizer industries.

Figure 6.3. A Resource Run – Phosphorous Example

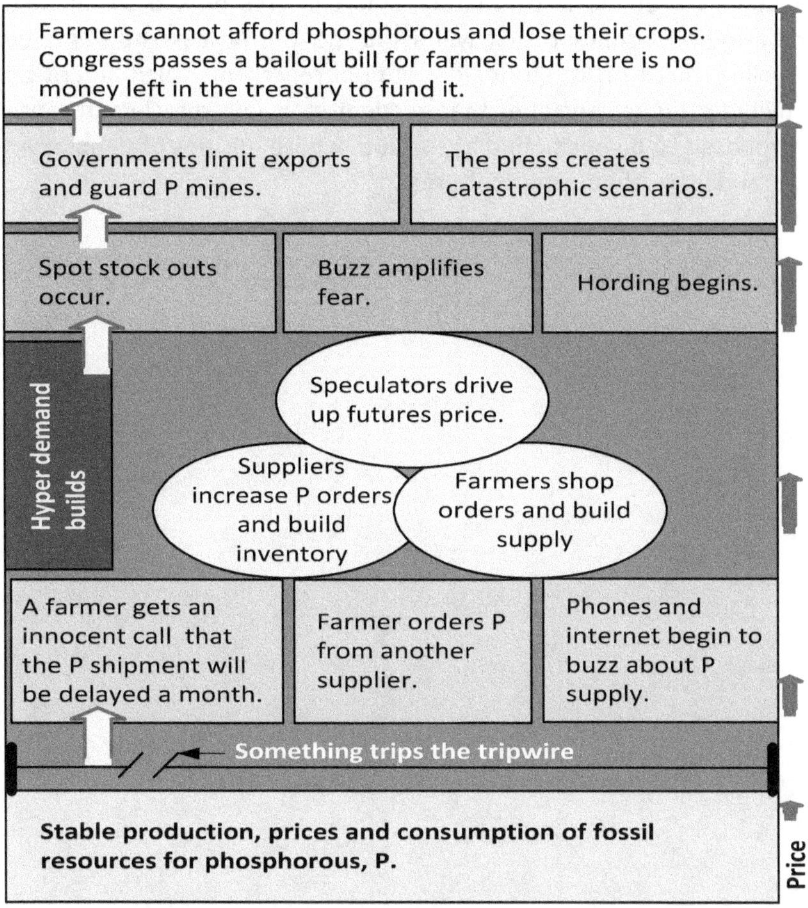

Farmers cannot afford phosphorous and lose their crops. Congress passes a bailout bill for farmers but there is no money left in the treasury to fund it.

| Governments limit exports and guard P mines. | The press creates catastrophic scenarios. |

| Spot stock outs occur. | Buzz amplifies fear. | Hording begins. |

Hyper demand builds

Speculators drive up futures price.

Suppliers increase P orders and build inventory

Farmers shop orders and build supply

| A farmer gets an innocent call that the P shipment will be delayed a month. | Farmer orders P from another supplier. | Phones and internet begin to buzz about P supply. |

Something trips the tripwire

Stable production, prices and consumption of fossil resources for phosphorous, P.

Price

The demand side of the food supply may create a food cascade where millions of people buy and hoard available food in anticipation of food shortages.[202] If everyone in a nation buys just one extra pound of flour at the same time, stores would have no way to restock their shelves. A food cascade could be ignited by a resource run, food price increases or shortages similar to the riots that jolted over 40 countries in 2008. The food riots may have served as a prequel to a food cascade. The consequence of a food cascade would be catastrophic.[203]

Smartcultures

The high probability of fossil fertilizer extinction by 2031 presents frightening scenarios. Fortunately, smartcultures offer a sustainable and affordable solution that will avoid the loss of fossil nutrients by recycling them. This strategic solution represents such a critical activity for human societies that nutrient recovery, recycle and reuse is proposed to be positioned at the zoo, where millions of people can view and learn about smartcultures.

Chapter 7. Nutrient Recovery with ZooPoo360

ZooPoo360
Recover, recycle and reuse energy, water and nutrients from the Zoo waste stream

Rather than position nutrient recovery on a farm were only a few people can see it, why not create a world class demonstration project at the zoo? If we can do ZooPoo at the Zoo, we can recover energy and nutrients anywhere.

Vision: A zoo becomes the world's first EcoZoo and demonstrates nature's way for energy storage and harvest – green solar energy in algae. Nature's first and simplest energy system, algae, uses only sunshine, wastewater and surplus CO_2 to recycle and reuse ZooPoo to produce clean, sustainable, carbon neutral food, feed, energy, fertilizer and freshwater.

ZooPoo360 enables the zoo to move towards a net zero:

- **Carbon footprint** – no net carbon dioxide emission.
- **Freshwater footprint** – no net freshwater consumption.
- **Fossil fuels footprint** – no net fossil fuels consumption.
- **Fossil nutrient footprint** – no net fossil nutrient consumption.

Smartcultures

Goals:

Create ZooPoo360 in a manner that enables the demonstration facility to become a destination eco-tour where guests learn about smartcultures and sustainable food and energy production systems. Target markets include especially those who currently pay to burn or bury their wastes and will benefit from learning how to recover their waste streams.

- **Farmers** who will be able to recover value from animal and plant wastes. Farmers will also be able to moderate pollution from runoff and clean polluted water.
- **Municipal waste facilities** that will be able to create value from human wastes.
- **Community (garden and trash) waste facilities** that will create value from organic waste streams including recyclable trash and garden clippings.
- **Power and cement plants** and manufacturers that will create value from their surplus CO2 while avoiding emissions.
- **Citizens** who desire to learn how to minimize their waste streams and ecological footprints.
- **Children** who wish convey the message of conservation and renewal to their peers and parents.
- **Churches** who desire to carry green and sustainable lifestyles to their communities.
- **Schools** who want engage students in ecologically sustainable systems.
- **Green and environmental social networks** who want to see ecologically responsible production of food, feed, fuels and freshwater.
- **Demonstrate and educate zoo visitors** in food and energy security by reclaiming and recycling surplus inputs that are affordable and unlike fossil resources, will not run out.

- **Transform a major zoo cost (ZooPoo) into a profit center** while demonstrating carbon neutral food, energy and freshwater production.

- **Demonstrate global stewardship** by moderating pollution while producing valuable products that store rather than release carbon. If a fraction' of people — 20% — made small, environmentally beneficial changes, such as switching furnace filters, using low energy light bulbs or driving less aggressively, overall energy consumption would drop by up to 20%."

- **Educate, demonstrate and diffuse** sustainable and affordable food and energy (SAFE) production that creates a positive ecological footprint.

- **Introduce smartcultures.** Enable visitors to see and learn the benefits of Sustainable Micro Algae Regenerative Technologies.

Challenge

Many animals, plants and entire ecosystems are threatened with extinction due to global climate change, food costs, freshwater scarcity and the availability and affordability of fossil fuels and fossil nutrients (fertilizers). ZooPoo360 addresses each of these challenges and shows visitors how to change their own behaviors to save the animals and plants and to enhance our communities. "Sustainable You at the Zoo" will provide a take-home checklist to support ecologically sustainable lifestyles.

Industrial food production is likely to crash by 2031 due to unavailable or unaffordable fossil inputs. Many food growing regions will be out of fresh water soon, others will run out of fossil fuels and farmers globally are likely to run out of P, copper and zinc within 20 years. Global warming, rising oceans, soil degradation and salt invasion put our entire food supply in severe jeopardy. The world needs a supplemental food supply that produces in spite of global warming and recovers recycles and reuses energy, water and nutrients from waste streams. ZooPoo360 provides a focal point for the critical actions that can save millions from starvation.

We can build an energy and nutrient recovery and recycling demonstration and learning facility that enable visitors of all ages to see, experience and learn how to adopt green behaviors and lifestyles. ZooPoo360 uses the Zoo waste stream, ZooPoo, and recovers and reuses energy and nutrients to produce electricity, freshwater, vitamins, minerals, health foods, animal feed, fertilizers and fine medicines. Algae recover the hydrocarbons stored in ZooPoo, which can then be used for energy rather than burning fossil fuels. The facility will also include demonstrations of other renewable forms of energy such as solar, wind and possibly geothermal systems.

The Zoo waste stream creates significant costs plus a large carbon footprint because the ZooPoo and plant wastes must be collected, stored, loaded, transported and buried in public waste dumps. Current actions forfeit the value of ZooPoo while waste disposal adds significant dollar and energy costs to the Zoo.

ZooPoo360 will demonstrate how to transform ZooPoo and recycle valuable food, feed, energy, freshwater and nutrients.

Background

Mother Nature has been producing algae for 3.5 billion years and ZooPoo360 builds on nature's first and most efficient food and energy production system. Most people are unaware of the potential for algae to provide carbon neutral food, feed and fuel because the few algae producers have been distant from population centers. ZooPoo360 raises algal farming to a new level by enabling field production of algae and its many coproducts from the Zoo's waste stream. After building the production / demonstration / education center, ZooPoo360 will be self-sustaining environmentally and economically.

Most of the energy on Earth comes from the sun. Nature transformed algal biomass from ancient oceans into fossil fuels but the process took 400 million years. Fossil fuel offer a convenient form of concentrated energy but pollutes our atmosphere with CO_2, other heat trapping gases as well as heavy metals and black soot particulates. Cultivating algae accomplishes in weeks what nature took eons to do – produce fuel – without fossil fuels and the associated pollution. Algal biofuels displace the use of fossil fuels and release only pure oxygen to the atmosphere. Algal oil creates clean, renewable biofuels that burn with no particulates because the algal oil has not fossilized; it is simple vegetable oil. After recovery of algal oil for energy, other coproducts may be extracted from the remaining biomass, especially food, feed and fertilizers

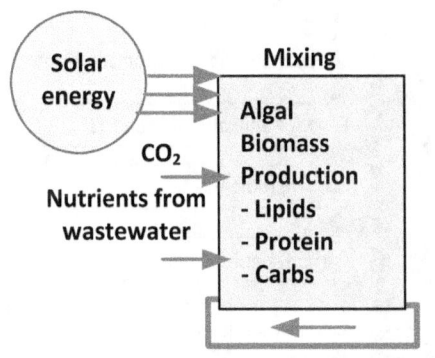

Recycle water and nutrients

Algae produce valuable biomass with surplus inputs that are cheap and will not run out – sunshine, CO_2 and nutrients from wastewater. Algal production systems get most the needed energy free from sunshine but require some additional energy for mixing and extraction.

Additional energy may come from pyrolysis of waste biomaterial, solar, wind or geothermal. The water used to culture the algae may come from any source but high-nutrient wastewater offers many advantages, especially free nutrients. Water may be recycled and the residual nutrients reused. Lipids are pressed out of the biomass for use as liquid transportation fuels. Separated protein provides food energy for animals, fowl and fish. The remaining carbohydrates can be refined into energy, biodegradable bioplastics, paper, fabrics and many other products. Algae can also provide vitamins, medicines and vaccines, Figure 7.1

Figure 7.1 ZooPoo360

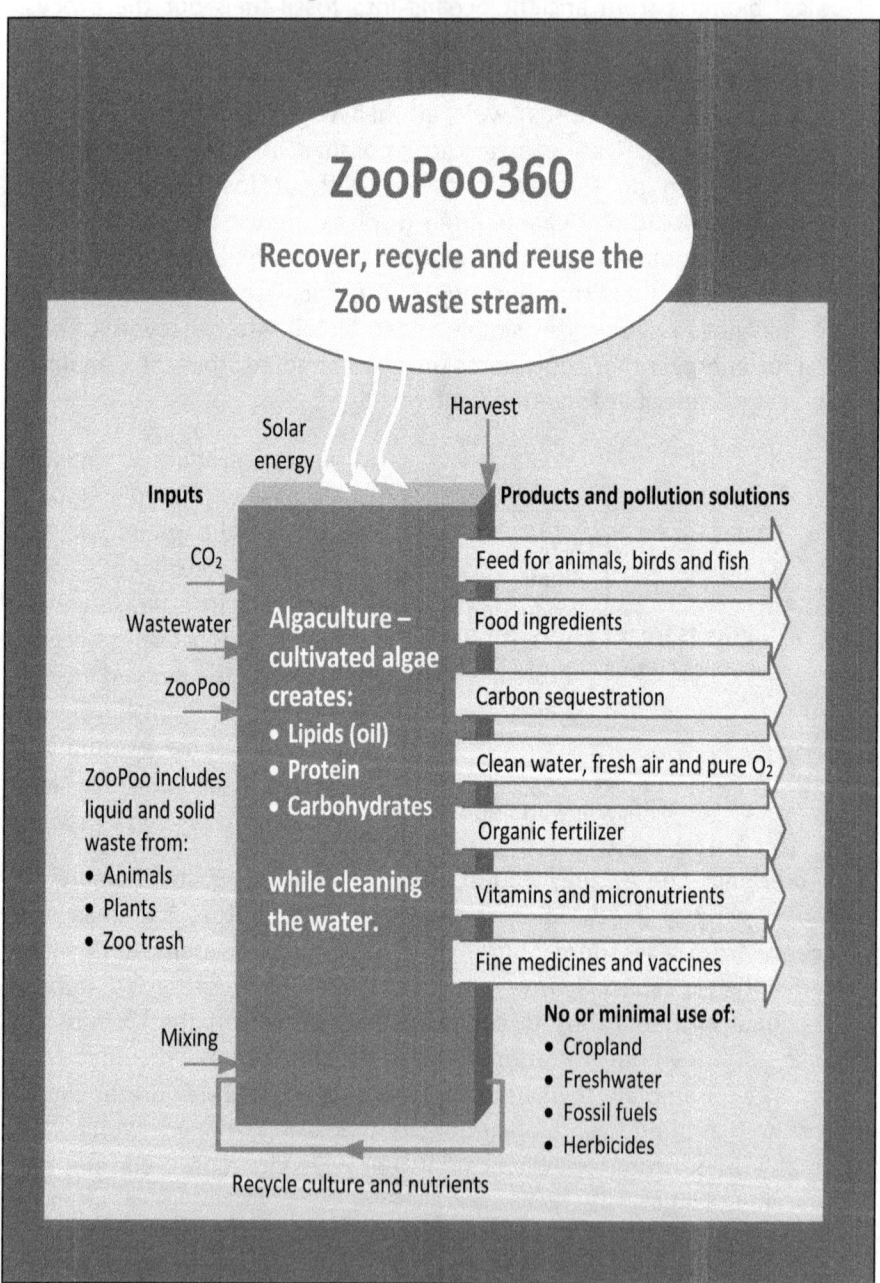

The ZooPoo360 process

The ZooPoo360 graphic shows that algae use free or surplus inputs while producing valuable products and pollution solutions.

ZooPoo includes the liquid and solid the wastes from animals, plants and zoo trash. ZooPoo is dumped into a pond where the nutrients are leached out. The biosolids are filtered, dried in the sun and then burned in a closed kiln in a process called pyrolysis. Burning the organic matter in the closed system releases no CO2 to the atmosphere and creates three valuable components: H_2 and CO gases and biochar. The H_2 is vented to a generator to create the electricity for the ZooPoo exhibit. The carbon monoxide gas is piped in an algae pond where it provides the carbon source for algae production. The biochar is sold to farmers as a slow release bile fertilizer and soil conditioner.

ZooPoo360 employs a clean and adaptable carbon neutral production process that consumes large amounts of CO_2 and transforms the carbon into high-value products while releasing pure oxygen into the atmosphere. The core technologies include:

- **Nutrient recovery** – algae bioaccumulate and store nutrients from the pootea created from animal and plant waste streams.
- **Energy recovery** – burning animal and plant solid wastes in a closed system, gasification or pyrolysis, creates H_2 for energy production and CO to feed algae.
- **Food and feed production** – algae are harvested and separated into component products with the protein going to produce feed for fish, birds and animals.
- **Clean water** – algae clean wastewater and make it suitable for animal or human use.
- **Energy production** – algae are harvested and the oil is pressed out to create clean, green diesel that burns with no black smoke particulates. Additional energy is created by the H_2 produced from gasification of solid wastes.
- **Fertilizer production** – selected residual from algae production may be used as a fertilizer for the many plants at the zoo. The

biochar, produced from pyrolysis, provides additional fertilizer and soil amendment for Zoo plants.

- **Fine medicines** – algae are harvested for various health foods, vitamins and minerals as well as Omega-3 fatty acids (found in fish oil and algae) which improve the health of zoo animals.

Carbon neutral production means that no new carbon enters the atmosphere.

Carbon neutral production

The ZooPoo algal production system illustrates the steps in carbon neutral production where fuel, electricity, food, feed, fertilizer and other products are made using no fossil fuels or fossil carbon products such as fertilizers or agricultural chemicals. The ZooPoo solid organic waste is dried in the sun and burned in a gasification unit, Figure 7.2.

Figure 7.2 Carbon neutral production

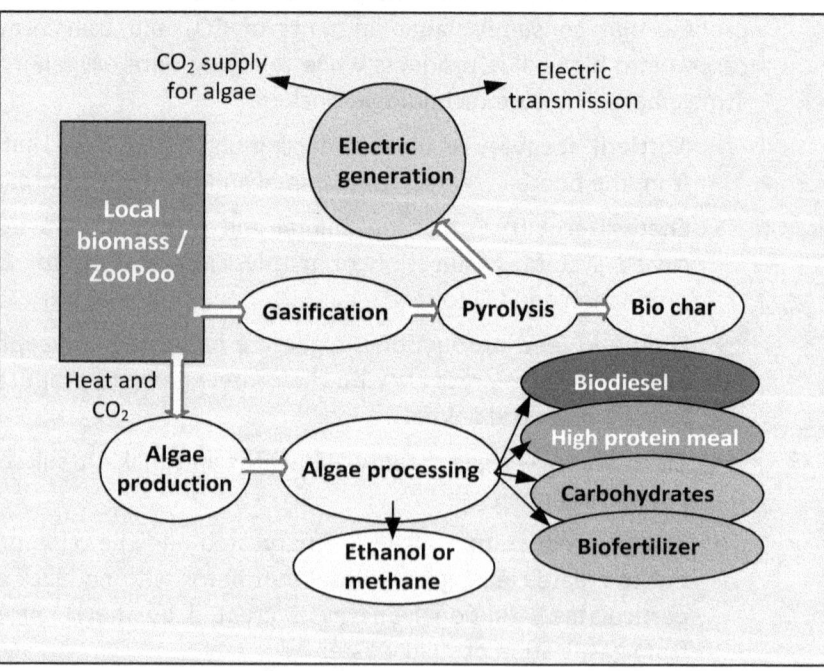

The gasification unit is a closed kiln that creates heat and carbon dioxide that provides a carbon source for the algae. Gasification and pyrolysis creates H_2 which is vented and burned to power an electric generator. The pyrolysis creates biochar which can be used in fields as a rich organic fertilizer and soil conditioner.

The algae may be harvested and processed to create biodiesel, high-protein feed, carbohydrates, biofertilizers and a host of other products. The carbohydrates can be converted to ethanol, methane, paper or textiles.

ZooPoo follows a set of values that provides policy guidance.

Values

1. We learn from and mimic nature's way, the Earth's oldest method of food, energy and feed production – algaculture.

2. We act as stewards of our Earth, sharing knowledge and technology globally so people can produce carbon negative food and energy locally. Carbon management represents a global challenge in order to mitigate climate chaos.

3. We focus on eco-plus production systems that are ecologically positive and improve the environment.

4. We take advantage of waste streams to recover, recycle and reuse precious energy and nutrients and we use active monitoring and metrics to insure the removal of all toxins, pharmaceuticals or heavy metals possibly captured from the waste streams.

We may not be pretty because waste streams are dirty but our products are clean for people, animals, plants and our environment.

Beneficiaries

ZooPoo360 will benefit zoo visitors of every age. The learning facility will serve as a gathering point for environmental and socially conscious networks. ZooPoo provides engaging learning opportunities for people interested in carbon neutral production of food and energy, ecologically sensitive lifestyles as well as water, food and energy conservation. Several groups will find particular value in seeing how we streams can be managed effectively.

- **Animals and plants**. Animals and plants will benefit from recycled freshwater, nutrients, food and feed.
- **Farmers**. Take advantage of agricultural waste streams to clean air, soils and water while recovering valuable energy and dissolved nutrients from moopoo, tweetpoo, fishpoo and fertilizer runoff.
- **CO_2 sources**. Take advantage of CO_2 pollution from fossil fuels that occur in power and cement plants, breweries or manufacturers to create valuable green biomass, to displace fossil fuels and to recycle carbon.
- **Municipal wastewater treatment**. Take advantage of human (peepoo) waste streams to clean air, soils and water while recovering valuable energy and dissolved nutrients.
- **Industrial wastewater treatment**. Take advantage industrial waste streams to clean air, soils and water while recovering valuable energy and dissolved nutrients. Industrial waste streams often contain heavy metals that algae removes and then must be separated from the green biomass using standard extraction technologies.
- **Transportation.** Provide sustainable and affordable liquid transportation fuels for planes, trains, trucks and ships that will continue to need liquid fuels for decades.

ZooPoo exhibit

The full ZooPoo exhibit will cover several acres and include ponds, cultivated algal production systems, (CAPS), the main exhibit and several satellite exhibits. The main exhibit will house:

1. Part of the CAPS (the other part may be behind the exhibit in the greenhouse)

2. The CAPS may use bags, tubes or flat plastic rectangles to grow algae.

3. An auditorium for 100 people with a projector and web connection for webinars.

4. An algal culture demonstration lab with 10 microscopes and 10 wet lab setups.

5. A small aquarium 10 x 30' in three sections showing tilapia growing at different stages of development.

6. A 20 by 30' greenhouse demonstrating aquaponics, growing vegetables and water enriched by tilapia urea.

7. An inviting entrance decorated with beautiful pictures of algae.

8. An exhibit showing how algae were used by ancient cultures.

9. An exhibit showing how farmers used mixed agriculture and organic production for centuries.

10. An exhibit showing how industrial farming uses P once and how P is lost to the field.

11. An exhibit showing how algae are cultured.

12. A room showing the many products made from algae.

13. An exhibit showing how ZooPoo can recover energy and nutrients from the zoo waste stream.

14. An exhibit showing how algae may be used for wastewater treatment and CO_2 sequestration. This exhibit may show how dairy wastes can be recovered, recycled and reused similar to ZooPoo. Ditto for municipal and industrial waste streams.

15. An exhibit showing how algal compounds provide medicines, nutraceuticals and vaccines.

Satellite exhibits

- **For kids** – How ZooPoo can be cleaned, recovered and reused for animal feed.

- **For bigger kids** – How ZooPoo can provide vital micronutrients and vitamins for animals.

- **How ZooPoo recovers wastes** and transforms them into valuable products.

What questions will ZooPoo address?

At the top level, ZooPoo will focus on the zoo waste stream to demonstrate how energy and nutrients can be recovered and reused for a wide variety of products and solutions. ZooPoo will demonstrate human future lifestyles in terms of sustainability, conservation, pollution solutions and ecological preservation. ZooPoo will provide practical demonstrations of solutions desperately needed by current societies including how to produce sustainable and affordable carbon neutral food, feed and fuel without using fossil resources.

Global warming has severe negative impacts the productivity and nutritional value of traditional, land-based foods. A few degrees hotter than normal can lead to a 30% loss in food production and a 15% loss in nutrient value. ZooPoo will demonstrate food, feed and fuel production that thrives with climate change and has neither production lost nor nutrient value loss with additional heat.

Rising ocean levels combined with sea salt invasion from tidal surges will destroy millions of acres of fertile cropland as well as displaced millions of people and animals that live on river deltas and coastal lowlands. ZooPoo will demonstrate food feed and fuel production that does not need fertile soils.

Global climate change creates warmer oceans which feed more severe storms, hurricanes and monsoons. **Severe storms** not only devastate crops but erode soils and pollute surface and groundwater. ZooPoo will demonstrate food, feed and fuel production that occur independent of climate and weather.

People globally are crowding into increasingly **dense city slums**. ZooPoo will demonstrate food and feed production from human and animal waste streams that can grow on rooftops, balconies, sides of buildings, vacant lots and over parking lots.

Modern diets filled with empty calories from refined genetically modified food grains such as corn and soy that create severe health impacts. Modern diets are plagued by monocultures. Jason Clay of the World Wildlife Fund reported that nearly 90% of our calories now come from only ten products.[204] The Centers for Disease Control report that one out of three Americans born after the year 2000 will develop diabetes, largely due to living on a cheap high fat, sweet and salty diet of largely empty calories. ZooPoo will demonstrate high protein, low calorie food production that is high in vitamins, minerals and micronutrients such as antioxidants.

Modern food production relies on **fossil inputs** that will be unavailable or unaffordable for farmers within the current generation. ZooPoo will demonstrate food, feed and fuel production built on a foundation of cheap and surplus inputs that will not run out, including sunshine, CO_2 and wastewater.

Farmers are failing because their cost for **herbicides, pesticides and fungicides** are increasing each year while resistant pests are increasing even faster. Global warming has accelerated the propagation of weeds, insects, fungi, molds and other predator vectors. ZooPoo will demonstrate food, feed and fuel production without costly agricultural chemicals.

Many **farmers are failing** because their input costs are too high and they have a high cost for managing the farm waste stream. ZooPoo will demonstrate how farmers can turn a substantial cost, the farm waste stream, into a profit center. Farmers will see how they can recover the energy and nutrients from animal and plant wastes as well as reclaim and recycle polluted water.

Our foods today often **travel thousands of miles** before they arrive at our local supermarkets. For many foods, transportation represents 50% of the cost. The continually increasing price of transportation fuels will push foods that require long transport beyond the means of most consumers. ZooPoo will demonstrate food production for locavores – people who prefer to eat locally produced foods. ZooPoo will show how a rich variety of great foods can be grown near any city including vegetable protein in algae, shellfish and finfish grown in aquaponics. Smartcultures using algae and fishpoo can provide the primary nutrient source and a wide variety of vegetables grown in hydroponically.

Interesting ZooPoo facts

ZooPoo includes liquid and solid wastes from animals and plants. Wastes such as paper products from Zoo concessions that end up in trash may also be recycled because they retain the nutrients from the plants from which they were made.

1. ZooPoo contains roughly 60% of the energy originally in the plants eaten by zoo animals. Elephant poo contains about 90% of the original plant energy because while elephants have the biggest appetites of all zoo animals, they have the lowest energy and nutrient absorption. Elephants are large, rich poo factories.

2. ZooPoo contains about 80% of the nutrients originally in the plants eaten by zoo animals. ZooPoo currently pollutes water, air and soils while creating manure that must be trucked for disposal in public landfills.

3. ZooPoo energy can be recovered from biowastes through gasification (pyrolysis) where the wastes are burned in a closed kiln creating H_2 and CO. The H_2 drives a generator creating electricity while the CO feeds the algae culture creating algal oil for energy, algal protein for feed and many other coproducts.

4. ZooPoo nutrients can be recovered in ponds where the poo is made into pootea and algae go to work and absorb the dissolved nutrients. Algae use solar energy to bioaccumulate nutrients and the end result is nutrient-rich algal biomass that can be recovered and reused. Algae also remove biowastes from wastewater and transform the wastes to clean, green plant biomass.

5. Zoo activities create a significant carbon footprint which algae mitigate by capturing two pounds of CO_2 in each pound of algae. When the algae are used for fuel or feed at the zoo, the CO_2 is recycled, creating carbon neutral production. Carbon

neutral fuel and feed production displaces the equivalent amount of fossil fuels that would otherwise have been consumed.

6. Zoo activities use significant amounts of freshwater that may be recycled and reused without fossil energy by using green solar energy captured in algae. Sunshine drives photosynthesis which cleans the water without fossil fuels.

7. Zoo activities use substantial fossil nutrients that may be recycled and reused without fossil energy by growing algae that recovers and recycles the nutrients using the power of the sun in photosynthesis.

A number of questions are obvious for a novel project like this.

FAQ

How to handle poo dominates the question set.

How will we handle the excrement question? We will turn poo upside down and celebrate poo as a treasure as we transform brown into gold; clean, green food and energy. We will use various steps in the process to assure cleanliness, healthiness and nutrient value. We will also use examples of how poo is used such as:

- China's farmers have used human poo on their fields successfully for thousands of years.
- Organic farmers use animal waste regularly on their fields to produce healthy crops.
- Beer and wine comes from the excrement of yeast cells.
- Millions of people cook their supper every night over dung.
- The Plains Indians as well as early settlers used buffalo poo regularly for their cooking fires.
- Poo retains high-value energy and nutrients, why waste it?

How will we make the process visually appealing? The demonstration facility will be beautiful with running water, great graphics and photographs (algae are the most beautiful and colorful plants on Earth) and wonderful smells. Algae emit only pure oxygen which smells like a redwood forest – without the trees. The terrible smell associated with algae ponds is not the algae but the bacteria that feed on algae and then die and decompose. We will avoid bacterial contamination for a host of reasons, most notably that we want to avoid predators such as bacteria whose favorite food is algae. We can avoid bacterial and predator invasion using a set of known parameters such as culture temperature, acidity and other factors.

How will we make the facility engaging for various ages and audiences? The ZooPoo facility will be a combination of beautiful art and science that engages people naturally. We will use some elements from the Arizona Science Center's Stroller Science® to engage the little people. We will create thematic and educational elements appropriate for each school level in order to support schools for field trips, internships and research papers. We will create waste to energy curriculums for various levels of education. Our challenge will not be guest engagement but avoiding information overload.

Each learning element will have a high-level summary of the U.S. and global situation defined in terms of impacts on animals and humans. The learning elements will include conservation management options for the global climate change and fossil resource extinction issues covered in the prior chapters as well as the following.

- **Animals.** What will be the plight of animals in 30 years?
- **Extinction** – Show pictures of the animals that will become extinct due to habitat loss, global warming and human impacts.
- **Threatened extinction** – Show pictures of the animals threatened with extinction due to habitat loss, global warming and human impacts.
- **Animal actions** – Identify practical lifestyle choices that guests can make in order to save animals from malnutrition, starvation and extinction.
- **Recognition and rewards** – Recognize that many animal, environment, botanical and social service organizations that are working to save our environment for animals and humans.

- **Engagement** – Identify opportunities for guests to become engaged with organizations locally whose actions can help reclaim our ecosystems and preserve animal health and welfare.

The learning center will be supplemented with a world-class website that enables real-time observation of the ZooPoo360 process and access to experiment and quality assurance monitors. School children, farmers and energy and food policy leaders globally will be able to do virtual tours locally and access learning modules.

How will we design the facility? We will assemble a consortium of experts to design the production system. The production system will illustrate how to use waste streams to produce sustainable and affordable food and energy. We will engage environmental design architectural students as well as professional architects to develop ideas for the learning facility. We will also rely on the Zoo Board of Directors to align the project with the vision and values embedded at the zoo.

How much space will be facility require? The ZooPoo facility will probably have a total land footprint of 5 to 7 acres. We expect to be growing algae in both ponds and vertical cultivated algal production systems, CAPS. The ponds will be about an acre each while the learning center will be on about half an acre. The learning center will be housed in a building where we will be producing algae in a close biofactory with a combination of sunshine and LED grow lights.

Staffing. The algal production facility will require several on-call professionals who may already be associated with the Zoo. The facility will need three operators who will staff the facility during zoo hours. The learning center and scientific demonstrations will be staffed with two professionals, several interns and volunteer docents.

How much will the facility cost? The algal production facility will cost roughly $7 million. This includes staffing and operations for three years, after which the business plan shows the zoo waste stream transforms to a revenue producing profit center. The production facility will cost about twice as much as a normal facility because we will be using a variety of solutions to minimize odor which will come primarily from the pootea, not from the algae. The learning center, including the algal art, scientific demonstration areas and learning lab with microscopes, will add another $10 million to the project.

The learning center will be able to host school field trips, disabled students and community groups in a manner that is both entertaining and educational. A projection system will enable guests to see various species of algae through powerful microscopes. The system will also enable the use of computer-assisted presentations and DVDs. The learning center will also be supported by an integrated, open source website designed to engage learners locally, nationally and globally. The website will enable anyone on Earth to do virtual ZooPoo360 tours and participate in learning sessions. The website will operate similar to Wikipedia so that experts can post summaries and papers on various aspects of conservation and sustainability.

Funding. We will develop a set of visuals as well as a business plan in order to attract funding from public and private donors. We will target donors that are not currently supporting the Zoo.

What are the risks of using CAPS? No unusual risks to crops occur with CAPS other than normal farming risks, e.g. weather, water and pests. We monitor and share publically over 50 parameters in order to ensure consistent high quality nutrient delivery and reliable crop production.

What if the algae mutated? Could it run wild and attack the field? No, algae mutate in nature but they make only very small changes. We are careful to use algae species during the growing season that provide valuable nutrients but do not create toxins. When we use algae that produce toxins for nematode management after harvest, we select non-indigenous species that will be crowded out by algae that have adapted to local conditions.

Will the CAPS become popular and then fade from use? The mass extinction of fossil resources will sustain the need to use crop inputs that do not rely on fossil resources. Algae's unique ability to recycle energy and nutrients from polluted water from waste streams will become more valuable as fossil resources become less available and less affordable.

How long will the CAPS system last? The system specifications are for 10 years. In hot, sunny exposures, the plastic plumbing may need to be replaced every five years.

How will the farmer know which species of algae to use? The field history, soil tests, geography, soil science, weather forecasts and intended crops will lead to recommendations of algal species and the timing of their delivery. Farmers will be able to use an interactive website to consider, select and modify nutrient options based on ambient conditions.

Can all of our fertilizer be delivered by CAPS? Current technology enables only partial fertilization plus the delivery of special micronutrients. Additional R&D may be able to develop full fertilization capacity in the future.

Are transgenic or genetically modified organisms (GMs) used? No, CAPS use only naturally occurring, indigenous Algal species because nature adapted algae over 3.5 billion years to produce precisely what plants need. We don't mess with Mother Nature.

Will algae grow on top of the soil and crowd out our targeted crops? No, field sites may find the topsoil crust green, orange or purple with algal biomass. However, algae do not compete with field crops for carbon or nutrients because algae have no roots. Crops may leach valuable nutrients from living filamentous algae or absorb nutrients as the algae decompose in the field.

If waste stream nutrients are used, aren't they dirty? We use current technologies adapted from food science to extract useful nutrients in a manner that is clean and safe. We also remove any heavy metals such as lead, arsenic, mercury and cadmium. The algal farm biofactory that prepares cultures monitors culture quality and cleanliness every hour. Samples from every batch sent to the field are retained for quality control analysis.

Chapter 8. How Does Nutrient Delivery Work?

We abuse land because we regard it as a commodity belonging to us. When we see land as a community to which we belong, we may begin to use it with love and respect. - **Aldo Leopold**, 1949, *A Sand County Almanac*

Nature provides very little for free and smartcultures are no exception. Smartcultures leverage micro biofactories that can capture N from the atmosphere and solubilize P and other nutrients locked in the soils. These micro biofactories can also bioaccumulate nutrients from waste, brine or ocean water. While these actions are not free, they can be accomplished with minimal or no consumption of fossil resources.

The nutrients needed to support crops must come from an organic or inorganic source. Smartcultures favor organic sources from waste streams but in circumstances where insufficient organic material is available, resort to inorganic chemicals to grow algal biofertilizers.

Farmers may choose to operate one or both nutrient recovery and/or delivery systems. A nutrient recovery operation with an accessible irrigation system may operate a single cultivated algal production system, (CAPS). The system grows large amounts of algae that capture nutrients from the waste stream and concentrates the culture. The recovery CAPS may harvest the algae to be spread on the fields or leave the algae in liquid concentrate. The concentrate may be spread directly on the fields or used to inoculate cultures in the nutrient-delivery CAPS near the fields. Farmers who do not have nutrient recovery systems may acquire their inoculants from universities, other farmers or private vendors.

Large farms may site their recovery CAPS on one part of the farm and place several delivery CAPS near the fields. Nutrient recovery CAPS are typically larger (1–10,000 gallons) while nutrient-delivery CAPS are smaller, (200–600 gallons). Another model uses one large CAPS and tanks on trailers are used to take the algae to various locations where it is metered into irrigation water or sprayed directly on fields.

Nutrient recovery CAPS use are solar energy, CO_2 and wastewater containing the leached nutrients to produce biofertilizer, plant growth hormones, soil conditioners and targeted nutrients such as extra calcium, iron or zinc. The system requires mixing and occasional quality control checks to ensure culture stability. The algae may be harvested, dried in the sun and stored like silage for later application to the fields. The concentrated liquid culture can be metered into the irrigation system or drained into sprayer for application to the fields, Figure 8.1.

Figure 8.1. Nutrient recovery CAPS

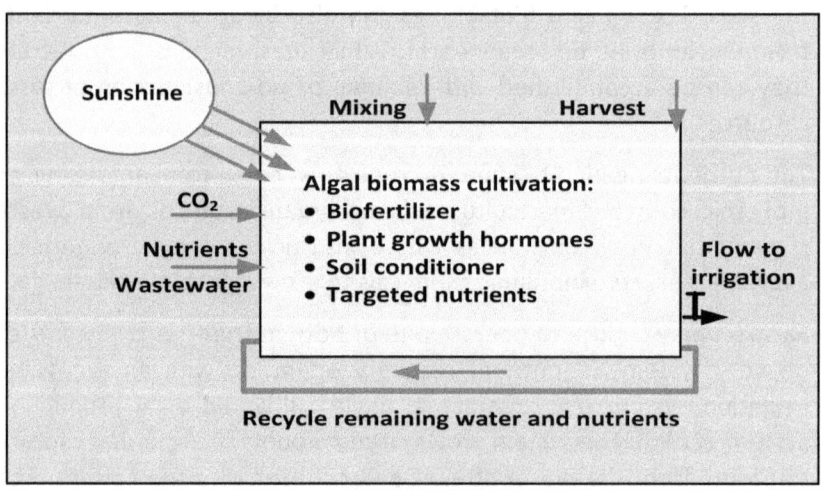

The configuration for nutrient recovery operations is site specific depending on the type of farm. An animal production facility may set up nutrient recovery near the manure source. Crop-based farms may site nutrient recovery at a convenient location on the farm. Smaller farms may operate as cooperatives for community nutrient recovery.

The smaller nutrient-delivery CAPS located near the fields supports 100 acres per 100 gallons of culture. Therefore, size may be determined by the configuration of the farm's irrigation system or field layout. Portable CAPS that sit on a small trailer reduce the number of systems required and provide considerable flexibility in supporting a variety of crops or fields planted at different times.

Figure 8.2. Nutrient-delivery CAPS near the Field

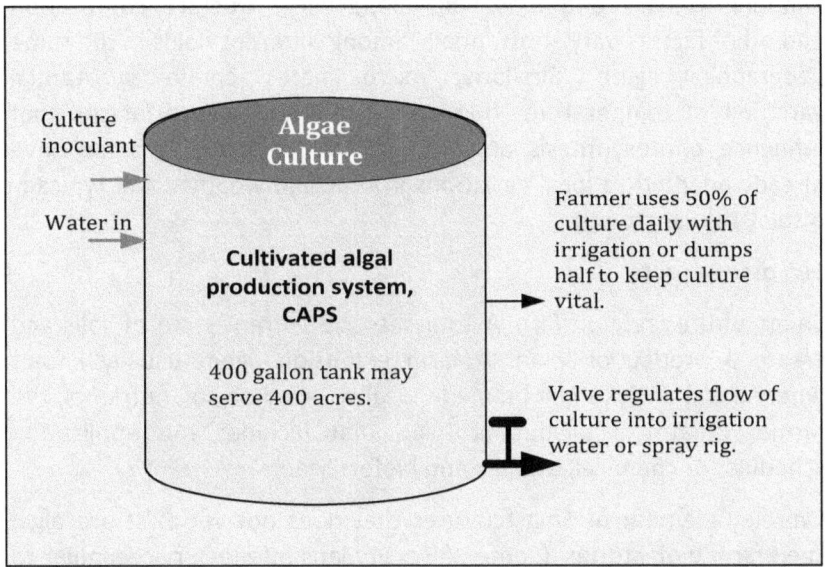

The water containing the algal culture flows with the irrigation water to the plants. Water delivery may be with flood, sprinklers or micro drip systems. Dry land farmers may use sprayers and some rice farms use aerial application. The fields may receive regular nutrient feedings during crop growth and development with different species used depending on the needs of the crop. Special nutrients such as calcium may be incorporated into the algal species for transport and uptake by the crops.

After harvest, an algal species with specially selected toxins may be applied to manage nematodes and other soil predators. The algae with toxins are naturally occurring and typically die out after killing the nematodes. While it is possible for algae to mutate, indigenous algae are far more robust and quickly crowd out any remaining algae that produce the nematodes toxins.

Biofertilizers produced locally may use algal strains collected locally or from known algal collections such as the University of Texas at Austin are the University of California at Berkeley. Research found that local algal species tended to outperform species from algal collections because indigenous algae have invested thousands of years in adapting to local conditions. Soil type, acidity, structure, compaction and other factors vary substantially among different fields in the same geographic region. Similarly, microclimates create substantial variation in temperature, humidity, winds and other factors that influence photosynthesis and crop growth. Indigenous algae have already adapted to local variations in soils and weather and typically display robust growth.

Soil assessment

Smartcultures rely on farmer to assess the current state of soils and create a production plan. A farm extension agent usually knows where a soil analysis can be made locally. Based on soil nutrients, the farmer creates a production plan that includes the application schedule for chemical, organic and biofertilizer.

One key element of Smartcultures that does not yet exist are algal seed farm biofactories. (Some university labs offer a service similar to seed farms.) Algal seed farms grow algal cultures in the laboratory from samples collected locally by farmers. Lab technicians separate various indigenous algal species from local samples and screen them for hardiness and the presence of desired compounds.

Once the specimens have been identified, pure cultures are grown to sufficient density that they can be sent back to farmers to inoculate their nutrient-delivery CAPS, Figure 6.3. A container containing the seed algal culture is delivered by courier to the farmer who puts the culture in the nutrient-delivery system. Algae are sufficiently robust that the culture remains viable for several days.

Once inoculated, the delivery algal culture grows to sufficient density in about five days. To use the CAPS with irrigation, the farmer sets the flow valve to use about half the culture daily. Since the algal culture doubles daily, a single unit may be used daily for the treatment of multiple fields. On days when no irrigation occurs, half the culture is drained from the system. The drained culture causes no pollution in the soils because it simply adds to the biomass of local algal species.

Figure 8.3. Seed Farm Biofactory

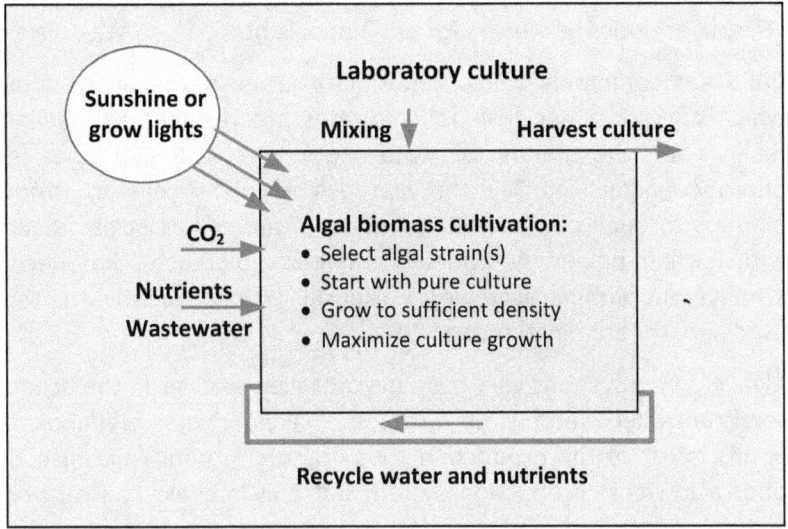

The University of Texas maintains a list of 3,000 living algal species that can be sorted based on characteristic, e.g. high algae oils. UTEX provides algae cultures at modest cost for research, teaching, biotechnology development and various other projects throughout the world. Their website lists the cultures maintained by UTEX, conditions for their long-term growth and information regarding the purchase of cultures.[205] The UTEX web site also offers specialty strains such as algae that grow in freshwater, extreme environments, snow and salt plains. Extreme environment strains come from tough settings such as Antarctica and the Gobi Desert. After strain selection, the next big decision analyzes the growing system design.

CAPS

CAPS are designed to capture maximum sunshine. Growing containers used for nutrient recovery and nutrient delivery may be open or covered ponds, plastic bags, plastic sheets, acrylics or glass – anything that allows light to penetrate. An algal seed farm culture is grown in a closed system, typically in a laboratory. Growing systems may be any size and use natural sunlight or artificial light.[206] Some systems use LEDs, fiber optics or mirrors for additional light.

CAPS are commonly called photobioreactors in the algal industry which our consumer research indicates creates fear and negative feelings in the minds of consumers. Even though the term photobioreactors implies the sun excites plant cells to produce biomass through photosynthesis, naïve observer's associate reactors with nuclear power. Additionally, the term bioreactor has become synonymous with garbage waste disposal.[207] Consequently, the terms used here for are biofactory or CAPS.

Algal growing systems vary from uncontrolled settings in the ocean to semi-controlled settings in estuaries, lakes, rivers, wetlands and ponds. Most of the production today occurs in ponds because they offer a low-cost production system but only modest control of the culture. Controlled and semi-controlled growing systems may be made of any material that allows light to pass. Containers must hold water and may be in any shape such as tubes, rectangles, barrels, blatters or bags. Hybrid systems start algae in the controlled environment of tanks and grow the production biomass in ponds.

Growers have traditionally used open ponds because they are easy to build and operate. Open ponds are inexpensive growing systems but allow algal species contamination, predator invasions, loss of water due to evaporation and other forms of contamination. Ponds enable opportunistic algal species to contaminate the pond and possibly dominate production based on growing conditions.[208] An algal pond may be infested with ciliates, amoeba or flagellates which can decimate the algal biomass within hours. An open algal pond loses about as much water due to evaporation as a grain field consumes in irrigation. Ponds may become contaminated with insects, chemicals, disease microbes, heavy metals or weed algae which degrade the culture.

Figure 8.4 Rectangular CAPS

Closed CAPS producers select algal species that maximize the characteristics desired such as biomass percentage of lipids, proteins, or component products. Food production would select to maximize biomass protein while biofuels may select a species with high lipid content.

Vertical, angled or horizontal? Algae are solar collectors so the plants benefit from maximum exposure to the sun. Some angled green solar systems track the sun similar to photovoltaic solar collectors.

Horizontal algaculture systems, typically tubular or plastic bags, provide another variation in solar exposure. Several companies use plastic bags that are typically rectangular or oval. XL Industries applies irrigation technology to algal production and their XL Trough uses a specially designed triangle trough system that rolls out in furrows behind a tractor. The XL Industries Super Trough system is simple and inexpensive.

Figure 8.5 Tubular CAPS

Rectangular or tubular? Different shapes provide different levels of solar exposure, Table 6.1. A wide rectangle, similar to an aquarium, holds a lot of water but does not allow each alga cell to have sun exposure very often. Consequently, thin rectangular tanks, about three inches thick, tend to out produce tanks that are wider. Tubular tanks may be a few inches wider because they present more surface area around the circumference. However, tubes around 6 inches typically out produce tubes that are wider.

Table 8.1 Cultivated Algae Production System Trade-offs

Type	Description	Limitations
Open pond	Economical, easy to manage, good for mass cultivation of algae, considerable global experience	Low culture control Stability issues Weak productivity High land use Species contamination Evaporation problems
Vertical column	High mass transfer, good mixing with low shear stress, low energy consumption, scalable	Small illumination surface Expensive construction Shear stress problems Cleaning issues
Flat rectangle	Large illumination surface area, good light path, good biomass productivities, relatively cheap, easy to clean, low oxygen build-up	Scale-up challenges Culture stability Temperature stability Possible shear stress
Tubular	Large illumination surface, good light path, relatively cheap	Gradients of pH, dissolved oxygen and CO_2 along tubes, fouling, high land use

The algal industry will continue to experiment with variations in CAPS shapes, sizes and with open and closed systems. It seems logical that low cost, low output systems will use open systems while those applications producing higher value products will use closed systems to maximize growth speed, vitality and species homogeneity.

Nature's challenges

Science must overcome a nontrivial set of nature's challenges to produce smartcultures reliably and sustainably. Farms that produce domesticated algae for specific purposes such as supporting modern agriculture, must leverage nature's strengths while overcoming some significant limitations. The major constraint to culture stability is called the law of the minimum, which refers to nutrient limitation, Table 8.2. When any important nutrient for algal growth becomes unavailable in the culture, because prior algal cells have already consumed the nutrient, algae stop reproducing and growth in the culture stops. Core nutrients include light, CO_2, N-P-K, sulfur and calcium as well as a variety of trace elements.

Table 8.2 Algal Growth Challenges

Challenge	Natural setting limitations
Law of the minimum	Algae grow quickly as long as the proper set of nutrients is available. When one nutrient becomes deficient, the entire algal culture stops growing.
Extremely limited light interchange	Sunshine or artificial light, the critical source of energy for algae production can be captured only from directly above in ponds, which significantly limits the number of photons that can be absorbed by algae.
Shading	Algae grow quickly in all directions but the cells that grow above prior cells tend to shade the older cells, which causes them to stop growing.
Contamination by other microorganisms	Other microorganisms that either compete with algae for nutrients or consume algae directly can significantly reduce culture growth and productivity.

Invasion by weed algae	Weed algae can enter a culture and propagate quickly to dominate the culture. Weed algae may substantially diminish the biomass value because it changes the composition of the algal biomass.

Algae grow so quickly that one or more nutrients are likely to be limited in the confines of a puddle, pool or pond. Since algae cannot migrate, the plants depend on nutrients coming to them and when those nutrients are deficient, the plant stops reproducing. Smartcultures production overcomes nutrient limits by assuring the culture receives a sufficient supply of macro and micro nutrients.

Algae typically absorb sunlight only from the tiny (few inch) interface between the atmosphere and surface of the water, which substantially limits its photosynthetic efficiency. A tree, for example, absorbs photons from hundreds of square feet of leaves that tower high above the ground. In natural settings, algae lack the vertical dimension which limits photon absorption. Photosynthetic efficiency can be increased by building vertical CAPS made of transparent materials that significantly increase solar exposure.

Shading creates another extreme limitation to growth in natural settings. As new plant cells grow, they tend to shade prior plants, which first reduces, then eliminates the solar energy necessary for photosynthesis. In natural settings, it is not unusual for algae to grow successfully only in the top two inches of the water where they can absorb sunlight. Shaded cells typically stop growing and may fall to the bottom of the water column. Others tend to drift in the water and are attacked by bacteria which consume the algae and create the unpleasant smell associated with stinky ponds. Producers minimize shading by continually mixing the algal culture. The same effect can be accomplished by growing algae in slowly running water or in the oceans where currents replenish nutrients such as in an upwelling.

Contamination from other microorganisms that feed on algae such as bacteria and rotifers enter natural algal stands and can propagate quickly and consume huge amounts of algae. These microorganisms may already be in the water, be blown in on the wind or be carried by insects or birds. Rotifers, like algae, can propagate exponentially and may consume 90% of an algal stand in one morning.

Weed algae present a similar problem. In nature, weed algae enter a pond and if conditions are right, may grow to dominate the culture. This presents no problem in nature because a diversity of algae species simply attracts a wide variety of hungry feeders. However, in domesticated solar gardens, weed algae are likely to change the biomass composition, similar to weeds in a traditional garden. A change from high protein to high carbohydrates may make the biomass less valuable to the community. Producers need to monitor and manage weed algae invasions which is accomplished by maintaining culture parameters such as temperature and pH.

Nature's limitations create a difficult but manageable set of challenges to domesticated algae production. Anyone serious about algae production should read the excellent source published by the Phycological Society of America, *Algal culturing techniques* edited by Robert Andersen.[209] A recent comprehensive report by algal industry scientists called the National Algal Biofuel Technology Roadmap published by DOE describes the full set of issues.[210] The Algal Biomass Organization recently released the Algal Industry Survey that prioritizes challenges in several areas and highlights recommended solutions.[211] Each of these sources is available at their web sites for free download.

Chapter 9. Forms of Fertilizer

The more we pour big machines, fuel, pesticides,
herbicides, fertilizer and chemicals into farming, the
more we knock out the mechanism that made it all work
in the first place.   **– David R. Brower**

Crop harvest and erosion work together to remove nutrients from soils, so they must be replaced with fertilizers. Fertilizer is the most widely used material in agriculture and comes in three types: chemical, organic and biofertilizer. Chemical fertilizers, also called mineral, synthetic and inorganic

Fertilizers are usually applied directly onto the soil, but can also be applied onto leaves (foliar feeding). Fertilizers can be either organic (e.g. manure or compost) or inorganic (mined or synthesized chemically). Organic fertilizers have been used for centuries whereas chemically-synthesized inorganic fertilizers were only developed in the 20th century. Fertilization can occur through biological processes like N_2 fixation or biofertilizers. Globally, mineral fertilizer is the major pathway for nutrient replacement because synthetic and mined fertilizers have been cheap, available and relatively easy to apply.[212]

Experiments with added N doubled grain production but the addition of N and P together produced five and six-fold increases.[213] Governments intervened directly to assure the supply of fertilizer to farmers through policies of price-fixing and subsidies. By the mid-1980s, subsidies for fertilizers reached 70% of the world price, pesticides 40% and water 90%.

Chemical, organic and biofertilizers

Each type has its disadvantages. The advantages of each fertilizer type need to be integrated in order to achieve optimum crop growth.

Table 9.1. Chemical, Organic and Biofertilizers

Advantages	Disadvantages
Chemical Fertilizers	
Typically cheaper and more available than organic fertilizers.	Diminishes produce taste and texture, shortens shelf life, decreases produce quality and reduces the total digestible nutrients.
Some nutrients are soluble and immediately available to the plants. The effect occurs very quickly.	Oversupply of N softens plant tissue resulting in crops that are more sensitive to disease, weeds and pests. Watery tissues also negatively affect taste and texture.
High in nutrient content so only modest amounts are required for crop growth.	Over-application results in negative effects such as leaching, water pollution, destruction of microorganisms and friendly insects, crop susceptibility to diseases, acidification or alkalization and reduction in soil fertility.
Light and far easier to apply than organics.	Adds toxic metals and minerals to produce.
	Reduces the colonization of plant roots with microbes. Inhibits natural symbiotic N fixation by

	rhizobia (soil bacteria) due to high N fertilization.
	Accelerates the decomposition of organic matter which degrades soil structure and promotes erosion.
	Nutrients are easily lost from soils through binding, leaching or gas emission, which diminishes fertilizer efficiency.

Organic fertilizers

Nutrient supply is more balanced which makes plants stronger and healthier.	Comparatively low in nutrients so larger volumes are needed to provide sufficient nutrients for crop growth.
Enhances soil biological activity, which improves nutrient mobilization from organic and chemical sources and decomposition of toxic substances.	Heavy to transport and apply. Transportation costs may make use impractical. Nutrient release rate may be too slow to meet crop growth requirements.
Enhances the colonization of mycorrhizae, which improves P supply.	Major plant nutrients may not exist in sufficient quantity to sustain maximum crop growth.
Enhances root growth due to better soil structure and encourage the growth of beneficial microorganisms.	Nutrient composition of compost is highly variable and in developed countries, the cost is high compared to chemical fertilizers.

Increases soil organics, therefore improving the exchange capacity of nutrients, improving water retention and buffering the soil against acidity, alkalinity, salinity, pesticides and toxic	Long-term or heavy application may result in salt, nutrient or heavy metal accumulation and may adversely affect plant growth, soil organisms, water quality and animal and human health.
Releases nutrients slowly and contribute to the residual pool of organic nutrients in the soil, reducing N and P leaching loss; also recycle minerals and micronutrients.	Manure must be plowed into the soil or the nitric oxides will volatize.
Helps to suppress certain plant diseases, soil borne diseases and parasites.	Animal manure may contain heavy metals + pharmaceuticals used on the animals.

Algal biofertilizers

All the advantages as organic fertilizers, plus:	May not be sufficiently effective on dry land farms. May need irrigation for nutrient delivery.
Improves soil porosity, cultivation cost, light and water penetration, moisture retention, seedling emergence and efficient gas exchange.	Displaces only up to: • 50% of chemical fertilizer • 30% of water use • 30% of fossil fuel use
Delivers full range of macro and micro nutrients.	May not be available or affordable locally.

Low cost delivery with irrigation or by sprayer.	Sparse research exists on growth stage specific nutrients.
Creates stronger plants that are more resistant to heat, pest, disease and drought stress.	Sparse field research except for rice.
Reduces need for: • Agricultural poisons. • Energy and fertilizers. • Cultivation	Novel and requires some change in farming methods.
Delivers special nutrients and hormones that stimulate plant growth.	

Fertilizer cost structures are likely to change with increases in fossil fuel prices and the coming scarcity of mined inorganic minerals and nutrients. Industrial fertilizers will become far more costly when communities begin to levy pollution and health taxes. Organic fertilizers will rise with the cost of fossil energy. Biofertilizer prices will decrease while they increase in nutrient value as we learn enhanced production methods, improved technologies for plant growth regulators and more about novel microbial tools for improving soil structure.

Algal biofertilizers

Algal biofertilizers (or inoculants) are a low cost, effective, environmental friendly and renewable source of plant nutrients that supplement and in some cases replace chemical fertilizers. Algae are photosynthetic organisms with chlorophyll that enables them to produce organic matter by photosynthesis (CO_2 fixation) and N_2 fixation. Microalgae live in symbiosis with lichens and mosses and make up cryptogamic crusts, which are often the major sources of biologically fixed N_2. Soil crusts act as a protective covering to minimize topsoil erosion from water and wind. Crusts may also provide the soil structure necessary for seed germination.

Algae are ubiquitous members of soil microflora and offer numerous advantages as biofertilizers. Algae do not compete with crops or other soil microflora for carbon since algae captures carbon along with N_2 from the air. Algae do not compete with crops for energy because neither can absorb more than a small fraction of the available sunlight. Fixed N_2 and other nutrients in algae become bioavailable to crops as a combination of leached N from living filaments and mineralization of decaying algal biomass.

Algae stimulate production of natural plant growth hormones that accelerate cell division and elongation, producing taller, greener and lusher plants that produce higher yields. Algae also stimulate plants to secrete compounds that repress harmful bacteria, fungi and other pests. In some cases, algae operate as a catalyst that helps plants manufacture natural insect repellent on their leaves.

Good soil consists of about 94% mineral and 3-7% bio-organic substances. The bio-organic parts are 85% humus, 10% roots and 5% edaphon. Microflora decomposes organic materials to produce humus. Edaphon may be the smallest soil component but plays a critical role for plant growth and development. Edaphon consists of microbes, algae, fungi, bacteria, earthworms, microfauna and macrofauna. Beneficial microorganisms operate to maintain the ecological balance by active participation in Nature's carbon, N, sulfur and P cycles. Soil microorganisms also play a pivotal role in building and enriching fertile soils. Algae also expand soils and increase porosity which makes room for the colonization of soil microbes.

Physical, chemical and biological factors influence the growth of algae on and in soils. Parameters include light, temperature, day length, nutrients, salinity, pH, atmospheric humidity, desiccation, wind, predators, pests, density and cycles of wetting and drying. Some species are intolerant to too much light and go dormant. Many algal species thrive and fix N_2 optimally between 30° and 35°C but are not active at higher or lower temperatures. In many cases, indigenous algae species can be found locally that have adapted to local light and temperature conditions and outperform pure laboratory strains.

Chemical properties are dominated by pH which influences species composition, growth and N_2 fixation. Algae grow best in neutral to alkaline soils. Nutrient and agrochemicals influence the activities in growth of algae and over application can be toxic. The addition of lime ($CaCO_3$) to fields stimulates algal growth.

Macroalgae biofertilizers

Most human societies that lived near coastlines harvested seaweed for human, animal and plant food because of its excellent nutrient profile. Macroalgae typically have vital minerals, vitamins and micronutrients such as iodine and vitamin A that are unavailable in sufficient quantities from local foods. Seaweeds act as natural fertilizing agents and stimulate helpful bacteria in the soil that release bound soil elements, fix N_2 and make nutrients available to the plants.

Considerable research has focused on biofertilizers derived from macroalgae (seaweed) such as kelp, Ascophylum Nodosum and fossilized kelp from calcium deposits which contain a broad spectrum of trace minerals. Some species of macroalgae have a slightly acidic pH of 5.5, created by the presence of amino acids, which enables it to help balance alkaline soils.

Seaweed biofertilizer and soil conditioners are made up of an array of water soluble minerals. While chemical fertilizers typically have only NPK nutrients, biofertilizers may have 75 minerals, growth hormones, cytokinin, auxins, vitamins and enzymes. Biofertilizers stimulate organic activity in the soil and lower toxic residues from various salts and chlorinated hydrocarbons. Toxins from harmful organisms such as nematodes and fungus infestations are also reduced by activated plant growth hormones.

Biofertilizers decrease the need for insecticides by supporting strong plants that are able to make a natural self-defense to insect invasions. For example, some fields treated with biofertilizers enable plants to produce a distasteful waxy film on their surfaces to repel insect attacks. Insects bypass treated fields in favor of untreated plants.

153

Algal biofertilizer improves the health and density of beneficial soil microbes which suppresses the proliferation of pathogenic microbes. Hardier plants build stronger cell walls due to the presence of additional silica (Si) which provides resistance to temperature spikes.

The research presented here used macroalgae or fossilized algae from calcium deposits harvested, processed, packaged and shipped to growers in a manner similar to traditional fertilizers. We believe the smartcultures model, where the grower produces algae locally for addition to irrigation water, will produce equal or better results. The preliminary research has been very encouraging.

Only NDAs, non-disclosure agreements prevent empirical results from being presented here. The intellectual property issues hold the algae industry hostage. The descriptions that follow are based on empirical research on algae biofertilizers.

The positive growth characteristics from biofertilizers are only partially attributable to algae itself. Many of the growth benefits occur because algae attract a diversity of beneficial soil microbes and microfauna that work in symbiosis to support plants. The microbes stimulate natural growth compounds that lead to accelerated growth and hardier plants that are more tolerant to stressors.

Biofertilizer research

Research on macroalgae biofertilizers compared with chemical fertilizers has not been peer reviewed, except where footnoted. Interviews with fertilizer producers, distributors and growers typically report significant (20% or higher) improvements in germination, plant density (due to higher germination rates), earlier maturation and larger, heavier yields with better taste. Biofertilizers reduce chemical fertilizer requirements and increase the plants' ability to withstand stresses from disease, insects, heat and drought. Both macro and microalgae typically deliver the full set of macro and micronutrients plants need. The results here are for macroalgae biofertilizer.

Rice. Biofertilizer improved germination rates, improved stands with higher yields, earlier maturation, larger heads, higher protein content and lower fertilization requirements.

Other grains. Biofertilizers improved germination rates, created better stands, larger heads and higher protein content. The increased bioavailability of vital nutrients has enabled crops to mature earlier and to withstand drought conditions better than untreated plants.

Corn. Biofertilizers supported increased germination rates, ear and kernel size, protein content, earlier maturation, increased yields both in silage crops and feed corn. Biofertilizers increased the crop's ability to withstand disease and insect infestation and increased the sugar content of the corn milk.

Hay. Hay grows faster, leafier, and had a faster recovery after cutting, lower water requirements, an increase in protein content and an increase in overall yields of 25% over the controls.

Cotton. Biofertilizers increased germination and growth rates, more blossoms, more squares, heavier setting of fruit with less loss dropping from the blossom to square to boll setting, sturdier stems and stocks and heavier setting of seeds in the boll. The increased luster to the cotton fiber increased the grade and the price. Bio-fertilized plants had lower nitrogen requirements, lower water requirements, higher yields per acre and increased disease resistance.

Tomatoes. Biofertilizer produced faster growth with larger, juicier, redder produce, lower acid content, improved flavor, an average of ten days to two weeks earlier maturing rate, a significant resistance to disease (principally the mosaic virus), increases in yields 10–23% in certain varieties (hot house conditions).

Citrus. Fertilized trees showed faster growth on young trees, a marked increase in sugar content of the fruit itself, thinner rinds, heavier fruits, higher disease resistance, lower fertilizer requirements, increased frost resistance (both to the tree and the fruit) and mineral deficiencies were far less prominent.

Fruit trees netted heavier yields of all fruits tested (including peaches, pears, plums, apples, apricots, nectarines, cherries), earlier maturation rates, heavier fruits, later fruits, better quality yields and lower fertilization requirements.

Sugar beets. Biofertilizer improved germination and growth rates, increased size and sugar content, improved disease resistance and lowered inorganic fertilization requirements.

Sugar cane displayed faster growth, earlier maturation, improved sugar content and cane quality as well as higher yields.

Melons showed higher germination rates, faster growth, higher sugar content, resistance to splitting and sunburning, earlier maturation, greater consistency in quality, lower water and fertilizer requirements, increased disease resistance and better quality retention (shelf life) after harvest.

Soybeans displayed an increase of 22% germination rate on 32 different experiments, 29% more nodulation in the rhizosphere, 21% yield increased, 9% protein increase, better disease resistance, lower requirements and earlier crop maturation.

Macroalgae biofertilizers offers significant value but have several downsides. Macroalgae supplies are typically mined from ancient oceans or harvested along coastlines. The limited sources could only supply a tiny percentage of the fertilizer needed for industrial food production. Harvesting, packaging and transporting macroalgae consumes substantial amounts of energy. Current manufacturers are not regulated by federal agencies which means growers are cannot depend on fertilizer quality.

Forms of Fertilizer

Microalgae biofertilizers

Most the research on microalgae biofertilizers has been performed in India and China. Of the 400 million farmers in the world, approximately 100 million farm in India. India has few fertilizer mines and recognized 80 years ago that fertilizers and agricultural chemicals are nonrenewable resources. India's scientist began exploring fertilizer production that avoided resource depletion and environmental degradation. Scientists recognized the impending energy crisis due to fast depleting mineral oil reserves and begin a search for N_2 production using biological processes that used nature's mimicry *in lieu* of the energy intensive Haber-Bosch protocol.

India's soil scientists recognized algal biofertilizers as a perpetual source of nutrients that did not contaminate groundwater or deplete fossil resources. Various biological systems capable of fixing atmospheric N_2 combine to contribute about 175 million tons of N to plants every year as compared to about 50 tons being fixed by industry.[214]

Biofertilizers are microorganisms which add, conserve and mobilize the crop nutrients in the soil and can lead sustainable crop production. The primary microorganisms used as biofertilizers belong to bacteria, blue-green (cyanobacteria) and green algae.

Biological N fixation represents an inexpensive source of N for increasing the productivity of crops. The biomass of these organisms decompose rapidly in soils and supply significant amounts of N-P-K, sulfur, zinc, iron, molybdenum and other micronutrients. The organic acids released during the biomass mineralization process accelerate P and micronutrient availability to crops. Biofertilizers amplify soil microbial populations such as bacteria, fungi and other microflora that activate soil enzymes which improve soil fertility.

India coined the term Algalization during the 1960s to describe the biofertilization of rice soils with free living blue-green algae. Algalization is a form of biofertilization where the fertilizer is grown in ponds, harvested and dried and then transferred to the field.

The government of India supplied a dried starter algal culture to Indian farmers who grew the algae for two months in small ponds. Dry algal flakes from the ponds, along with additional P and Fe were added to rice fields one week after transplanting rice seedlings. Algalization provided about half of the N needed for crop growth and development. Grain yields increased 33% with added inorganic N and about 16% with no added inorganic N.

Biofertilizer inoculation

Biofertilizers are generally applied to seeds, seedlings or soils but are unlikely to have a positive impact on plants unless they are able to grow and multiply. The introduced microorganisms will decline within days or weeks unless soil conditions support their growth. Inocula formulation and application methods are critical for biofertilizers. Effective inoculation management includes concerns about shelf life, suitable carrier materials, susceptibility to high or low temperatures or humidity, problems in transportation, storage and application.

Seed inoculation applies specific microbes that can grow in symbiotic association with plant roots. Soil conditions must be favorable for the inoculants to perform well. Selected strains of N-fixing Rhizobium bacteria have proven to be effective as seed inoculants for legumes. Seed inoculation may occur with multiple types of microorganisms. Seed treatment with Rhizobium, Azotobacter or Azospirillum, coats the seeds and allows them to dry. Each seed has several layers of the material as the inoculants treat the outer seed layer. The layering process maximizes bacteria to generate better results.

Soil inoculation

Microbes added to the soil compete with microbes already living in the soil that are already adapted to local conditions and greatly outnumber the inocula. Inoculants of mixed cultures of beneficial microorganisms have considerable potential for controlling the soil microbiological equilibrium and providing a more favorable environment for plant growth and protection. Joint inoculation of biofertilizer with mycorrhizae and N_2-fixing bacteria has been successful.

Microflora and their functions as biofertilizers

Microflora includes a wide variety of microorganisms such as fungi, yeasts, bacteria and algae. Some of the prominent microflora and biofertilizers found in soils follows.

Azolla. Azolla is a fern visible to the naked eye that lives in symbiosis with free living cyanobacteria that fix N2. Azolla is used in India, the Pacific Rim countries, West Africa and Brazil to provide organic N2 to rice paddies. Azolla significantly increases soil microbial and enzyme activity which improves soil fertility. When used as a green manure in rice paddies, Azolla has reduced the need for chemical N fertilizer by 30%. Azolla used as a biofertilizer increased rice yields 35--50%.[215]

In spite of the production increases low cost, low labor N fertilizer has significantly reduced the use of Azolla. Other problems minimize the use of Azolla. Production is labor intensive and there may not be room for growing both Azolla and the rice crop. For optimal growth with N_2 uptake by the rice plants, Azolla needs to be plowed into the soil and the field typically needs additional P and Fe fertilizer.

Rhizobium. Rhizobia are symbiotic bacteria that fix atmospheric N_2 in plant root nodules for certain crops such as legumes. A legume plant that grows N fixing nodules meets not only its own N requirement but enriches the soil in content for the following crops. The plant roots supply essential minerals and newly synthesized substances to the bacteria. Rhizobium inoculation ensures adequate N supply for legumes in place of N fertilizer and can fix 50–300 kg N/ha. Inoculation improved bean and pea crop yields of dry matter and crude protein by 54%.[216]

Azotobacters and azospirillum. Azotobacters and azospirillum are free-living bacteria that fix atmospheric nitrogen in cereal crops without symbiosis. They do not need a specific host plant. Azotobacters are abundant in well-drained, neutral soil and can fix 15–20 kg/ha N per year. Azotobacter sp. produces antifungal compounds to fight against many plant pathogens. They also increase germination rates and vigor in young plants which create denser crops. The bacterium produces abundant slime, which helps in soil aggregation and provides organic matter.

Plants, trees and vegetables benefit from Azotobacter. Cereals, vegetables, fruits, trees, sugarcane, cotton, grapes, banana, etc. are known to get addition N requirements from Azotobacter. Azotobacter inoculated seeds may increase germination by 20–30%. Application recommendations are as follows.

- For seed treatment use Azotobacter-biofertilizer at the rate of 250 g biofertilizer for 10–15 kg.
- Seedlings inoculation uses 4–5 packets (2–2.5 kg) per acre.
- For fruit crops, sugarcane, and trees it is used as soil application. At the time of planting fruit tree 20 g of biofertilizer mixed with compost is to be added per sapling.

Research reviews show that Azotobacter- and azospirillum-inoculated seeds or fields improve maize (corn), sorghum, cotton, rice, sugarcane, tomatoes, vegetables and oil seed crops from 9–72%.[217] Unfortunately, a wide variety of variables act in combination to create substantial variation in crop improvement. Azotobacter is the best-selling microbial product in India because farmers can see improvement in plant vigor in the early stages of growth.

Azospirillum colonizes in the root mass and fixes nitrogen in an environment of low oxygen. These organisms have low energy requirements and are effective even in saline alkali soils. They perform well up to 30–40° C, and establish well in the plant roots.

Azospirillum increases grain yield significantly and has been found to be especially effective in rice and cereals like wheat, barley, oats and pearl millet. Azospirillum creates visible crop improvement due to its ability to enhance the biomass of the root system, thereby allowing greater surface area for absorption of native nutrients.

Blue-green algae – cyanobacteria. Nearly all crop growing regions have indigenous blue-green algae, both free living and living in symbiosis with plants in the soil. These microorganisms have forms such as Anabaena, Nostoc and Calothrix and are especially present in rice growing regions because they have been colonized to support rice production for hundreds of years. The use of blue-green algae in India termed algalization in the 1960's, has been found to supplement N fertilizers about 30–40 kg N/ha/season.[218]

The growth of blue-green algae in saline, alkaline habitats reduces salinity by 25–30%, improves pH, electrical conductivity and exchangeable sodium. Some projects isolate pH-specific algal forms from the natural algal flora, which can be used in acidic soils without any soil amendments.[219] Natural indigenous algae typically outperform selected strains from other areas probably because they have adapted to local soil and climate conditions.

Phosphates solubilizing bacteria, (PSB). Under acidic or calcareous soil conditions, large amounts of P are fixed in the soil but are unavailable to the plants. Phosphobacterins, mainly bacteria, algae and fungi, can make insoluble P available to the plant. The solubilization effect of phosphobacterins comes from the production of organic acids that lower the soil pH and bring about the dissolution of bound forms of phosphate.[220]

PSB culture increased yields up to 200–500 kg/ha, saving 30–50 kg of superphosphate per hectare (about 50%). Phosphobacteria increase the number of root nodules, shoot length, root length and pod yield. The response to phosphobacteria operates most effectively in soils rich in organic matter and low in available phosphorus. Microphos containing *Pseudomonas striata* and *Bacillus polymyxa* increased the yield of wheat, rice, chickpea, sugarcane and potato significantly.[221]

Currently, P solubilizers are cultured by agricultural universities and private enterprises and sold to farmers through governmental agencies. Unfortunately, PSBs are marketed in India with few quality control checks, which led to very mixed results.

Vesicular arbuscular mycorrhiza, (VAM). Mycorrhizae are symbiotic relationships between fungi and plant roots. VAM fungi spread inside the root growing structures known as vesicles and arbuscules. The plant roots secrete substances to the fungi and the fungi benefit the plant by transmitting nutrients and water to the plant roots. The fungal hyphae may extend the root width, diameter and length 100 times.[222]

The hyphae enable the plant to reach into additional and wetter soil areas and help plants absorb nutrients, especially commonly limited mineral nutrients such as P, zinc, molybdenum and copper. Some VAM fungi form a sheath around the root providing a protective cover. Mycorrhizae enhance seedling tolerance to drought, high temperatures, infection by disease fungi and soil acidity. Application of VAM produces better root systems which combat root rotting and soil borne pathogens.[223]

Plant growth promoting rhizobacteria (PGPR). PGPRs represent a wide variety of soil bacteria that grow in symbiosis with a host plant and stimulate host growth. PGPR modes include fixing N_2, increasing the availability of nutrients in the rhizosphere, positively influencing root growth and strength. PGPRs have been found to increase plant biomass and nutrient content.

Biofertilizer limitations

In their review of biofertilizers, Wani and Lee concluded that dominate characteristic common to most biofertilizers in India lies in the unpredictability of their performance.[224] Crop responses to biofertilizers are not as dramatic as those with chemical fertilizers. These biological agents are subjected to a spectrum of hostile factors and their survival and efficiency is governed by variables such as the host plant, soil fertility, organics, acidity, moisture, cropping practices as well as biological and environmental factors. The effectiveness of biofertilizers also depends on the presence favorable microorganisms.

Biofertilizer adoption in India has been undermined by the poor quality of the inoculants. Farmers need better knowledge about inoculation technology and effective inoculant delivery and supply systems. During the 70s and 80s, many small biofertilizer producers in India began selling poor quality inoculants to farmers who lost faith in biofertilizers.

Forms of Fertilizer

Summary

Globally, farmers face an acute need to find cheaper sources of renewable plant nutrients. Sustainable food production also requires fertilizer application that repairs ecosystems rather than damaging them. Many organic farming advocates frame their arguments that chemical fertilizers destroy soil structure and the application of organic fertilizers can recoup the loss of soil tilth. Others advocate a judicious combination of chemical and organic inputs to meet the shortfall of chemical inputs. We may need the appropriate combination of chemical fertilizers, organic manures, crop residues, composting and biofertilizers that are affordable to farmers and enable sustainable food production.

The critical question is whether there is enough organic matter and microbial inputs available for intensive farming. The supply question has yet to be resolved but biofertilizers can play a critical role in both affordability and sustainability of the inputs needed for food production. Algal biofertilizers that recycle farm wastes offer a cost-effective method for replacing extracted nutrients, rebuilding extracted or eroded organic matter and enhancing soil structure.

Chapter 10. How should we Address World Hunger?

To cherish what remains of the Earth and to foster its renewal is our only legitimate hope of survival.
– Wendell Berry

Smartcultures expands the discussion about sustainable and affordable food because humanity faces global climate change and the impending mass extinction of fossil resources. The challenge of growing sufficient sustainable food to feed our hungry world is far too important to rely on any one approach or technology. Smartcultures offer only a partial solution. All reasonable alternatives need to be considered. Recent dialogues have made impassioned pleas to support hungry nations with food gifts or the inputs that would enable farmers to grow their own food. Other debates have centered on high-tech solutions, primarily transgenic crops and even low-tech solutions such as organic farming.

Gifting Food?

In *Ending Hunger Now,* former Senator, Presidential candidate and United Nations ambassador on hunger George McGovern suggests that the U.S. and other wealthy countries give more food aid to countries in need. He notes that the cost of hunger is unacceptable. "Today's malnourished pregnant and nursing mothers are producing tomorrow's barriers to personal, social and economic development – malnourished, brain dulled, listless children. Those fortunate enough to survive will dry through uncertain light, permanently diminished, unable to be productive, happy human beings."[225]

Senator McGovern is absolutely right about the unacceptable cost of poverty but gifting food is unsustainable and creates dependence. Wealthy countries can and have gifted food in the short-term but soon will have insufficient food to give. No nation has the underlying fossil resources to sustain large food gifts or, in the near future, even the expense of food transportation.

The U.S. policy of providing cheap food for citizens, heavily subsidized food for biofuel feedstock and gifting food as foreign aid has turned out to be extremely expensive, unsustainable and damaging to food recipients. The politics of cheap food and biofuel feedstock comes easily when one third of the food production costs are government subsidized and another third (ecological and human and animal health impacts) are ignored, with no accounting. Many farmers benefit from subsidized fossil fuels, freshwater, electrical power and commodities such as corn.[226] Farmers pay no cost for the resource consumption or ecological damage from food production. Decades of ignoring these costs will leave our children without critical natural resources because food accounting fails to consider:

- Loss of nonrenewable freshwater – 1000 tons of freshwater per ton of grain and about 43,000 tons of water per ton of beef.[227]
- Huge subsidies to big oil that discount fuel to farmers.
- Air pollution and health impacts from growing crops – 2.25 tons of CO_2 per acre, plus nitric oxide.[228]
- Nutrient loss from crop harvest and erosion.
- Nutrient erosion in modern produce with insufficient nutrients.
- Water pollution from fertilizers, pesticides and herbicides.
- Dead zones in rivers, lakes, bays, estuaries and oceans.
- Soil erosion – about six tons lost soil per acre per year.[229]

The U.S. foreign-policy operates independent of natural resource conservation, which threatens to drain America's bread basket dry within two decades. William Ashworth reported in *Ogallala Blue* that

Midwest farmers mine five trillion gallons of fossil water a year from the Heartland's aquifer that covers eight states from North Dakota to Texas. The Ogallala aquifer contains water laid down a millennia ago and is not replenished by annual rains. Children of Midwestern farmers will not inherit the valuable farmlands they expect but only dry prairie land that will be practically worthless. They will not be able to grow crops because all the irrigation water was mined by prior generations.

Gifting food provides only a very short term ease of hunger. When the gifted food is gone, people become hungry again quickly. Gifting subsidized food undermines local production. Even before the tragic earthquake, Haiti had lost nearly all its farmers because cheap subsidized American corn forced Haitian farmers to abandon their land. The farmers could not produce food as cheaply as subsidized American grains. The U.S. subsidies have forced 1.5 million Mexican farmers from their land for the same reason.[230] British International Development Secretary Douglas Alexander said in 2008: "It's unacceptable that rich countries still subsidize farming at $1 billion a day, depriving poor farmers in developing countries $100 billion a year in lost income."[231]

Gifting food inputs?

Professor Jeffrey Sachs of Columbia University and head of the UN Millennium Project, published *The End of Poverty: Economic Possibilities for Our Time* that provides a blueprint for cutting hunger and poverty. He urges industrial nations to double aid to poor countries from the current level of one quarter of 1% to one half of 1% of national income.[232] Professor Sachs makes a strong argument that wealthy countries should give poor farmers the inputs for food production, the training to produce, education, school meals and medical assistance.

Gifting the inputs to grow food seems like a great idea, except for the math. Unfortunately, the escalating cost and decreasing availability of fossil resources make gifting food inputs impractical. If India cannot afford to subsidize its own farmers for a single input – fertilizer –

wealthy countries cannot possibly supply the full range of fossil resources needed to support indigenous farmers.

The Millennium Project may be possible if a new sustainable and affordable food production technology were deployable that used few external inputs. A distributed, low external input agricultural system with affordable inputs to all farmers would zero out transportation costs and enable farmers on all continents to grow the food they need for their families and community.[233]

Transgenic crops?

Many people espouse GM seeds because genetic improvements formed the basis for the prior Green Agricultural Revolution. GM seeds may be necessary for meeting the global food demand but they pose several problems: equity, erosion, energy and ecological degradation. Transgenic seeds are no longer affordable to farmers in many countries, including many small farmers in America. GM seeds accelerate erosion because the seeds need extensive cultivation that compacts soil, decimates beneficial microorganisms and makes soil vulnerable to erosion.[234]

Transgenic crops produce higher yields but cannot compete with natural grasses – weeds. Proof of their inability to compete with natural plants can be seen in the second year when a field swarms with weeds rather than volunteers of the transgenic crop. Consequently, farmers must till the soil before planting to remove weeds that would compete for soil moisture and nutrients. GM crops consume more energy because they receive additional cultivation and herbicides to control competing weeds.

Highly productive seeds are designed for planting two or three times closer than traditional crops. Dense plantings consume substantially more water which means many dryland farmers have to put in irrigation systems. Dense plantings also consume significantly more fertilizer which also creates more soil, air and water pollution. Growing crops in crowded rows diminishes root development, degrades soil structure and yields produce without valuable micronutrients – vitamins, minerals and antioxidants.

Transgenic seeds may offer modest relief in the sense that plants may be able to germinate in drier and saltier soils. However, little is known about the GM drought or salt tolerant seeds because they are still several years away and have not been tested in various growing situations. Transgenic seeds are often more vulnerable to pests, weeds and disease, which may offset other benefits.

Genetic engineering studies have shown that changing a target gene, to improve N absorption efficiency for example, also changes several unrelated genes that help protect the plant against disease. A gene change designed to improve tobacco's N efficiency changed the plants toxicological properties.[235] Imagine what properties may change with genes that alter drought, heat, pH or salt tolerance.

Many farmers, including those in the European Union find their communities and social belief systems forbid the use of "unnatural" seeds. Therefore, natural processes such as biofertilizers will be necessary to provide sufficient affordable foods that align with the social belief system.

The cost of transgenic seeds will magnify social injustice, the difference between rich and poor farmers, communities and countries. The cost escalation of transgenic seeds occurs because a few companies practice oligopoly pricing.

Three companies – BASF of Germany, Syngenta of Switzerland and Monsanto of St. Louis – have filed applications to control nearly two-thirds of the climate-related gene families submitted to patent offices worldwide. These "climate ready" genes will help crops survive drought, flooding, saltwater incursions, high temperatures and increased ultraviolet radiation – all of which are predicted to undermine food security in coming decades.[236]

Company officials deny the climate-ready seed applications amount to an intellectual property grab. They say GM seeds will be crucial to solving world hunger but would not be developed without patent protections. Monsanto currently makes 60% of its revenue from genetically modified seeds.[237]

Some Monsanto seeds doubled in price in 2008, upsetting farmers who must buy new seeds from the company every year. Monsanto has sued numerous farmers for using seeds the company said contained their patented genetic material. Critics such as competitor Pioneer Seeds calls Monsanto's behavior a "platform monopoly" that crushes competitors much like Microsoft's Window's platform.

The *New York Times* reported that the Justice Department's antitrust division is investigating Monsanto for anticompetitive practices in the seed industry. The Justice Department is investigating whether Monsanto unfairly used genetic licenses to dominate the engineered seed market because 93% of soybean and 80% of corn plantings in the U.S. in 2008 contained Monsanto's Roundup Ready trait.[238] The Roundup Ready 1 trait patent expires in 2104 and Monsanto is forcing other seed suppliers to use their new Roundup Ready 2 trait in seeds that will effectively extend Monsanto's patent protection.

Britain's Soil Association study concluded that U.S. GM crops have been an economic disaster which has caused some farm groups to call for a moratorium on GM wheat, the next proposed crop to altered. The study estimated that gene-altered corn, soy and rapeseed cost the U.S. economy $12 billion since 1999 in farm subsidies, lower crop prices, loss of major export orders and product recalls.[239]

The biotechnology industry promised to solve the challenges of climate change and feeding the burgeoning world population while reducing agriculture's chemical impact. A 2009 study sponsored by the Union of Concerned Scientists examined the impacts of GM crops on pesticide use in the U.S. and found a dramatic rise in the use of herbicides on genetically engineered crops.[240] Charles Benbrook determined that 383 million additional pounds of herbicides have been used on GE crops since 1996, over what likely would have been used if GE crops had been replaced by conventional, non-GE varieties. The report shows the overall chemical footprint for engineered crops is immense and expanding.[241] The growth in herbicide use has important implications for public health, the environment and farmers' bottom lines.

The most contentious issue with GM crops is their impact on human and animal health. Everyone associated with GM crops knows that moving or adding a gene in a complex DNA sequence can have multiple unanticipated impacts on the organism. Monsanto did the R&D on their GM seeds and reported no adverse impact on mice feeding trials. For years, Monsanto held the research data privately, claiming intellectual property requirements. The Union of Concerned Scientists recently won a judicial judgment to have the results examined by a scientific third party. Unsurprisingly, the independent examiner found that the GM seeds did impact the rodents' kidneys (though how much remains under debate). New tests are underway with longer feeding trials and stronger controls on key variables.

Independent of GM seeds, many people assume technology will rescue the world from the brink of mass starvation.

Technology rescue?

Sustainable food production may not be possible with traditional land-based crops unless breakthroughs reverse current actions and major innovations occur in several areas:

Table 8.1 Innovations Needed for Crop Production

Innovation	Crop inputs to produce food
Nutrient efficient	Minimizes fertilizer application, nutrient waste and pollution.
Water light	Requires minimal or no fresh water.
Energy light	Requires minimal or no fossil energy.
Fertilizer light	Requires or no added chemical fertilizers
Chemical light	Requires minimal or no agricultural chemicals.

Crop outputs – productivity	
High yields	Grows high yields of good food on small acreage that diminishes the demand for cropland.
High quality	Grows produce with great taste, texture and color.
High nutrient density	Grows produce with high density of macro and micronutrients as well as vitamins and minerals.
Harvest waste	Harvestable with minimal waste.
Climate hardy	
Robust	Tolerant to crop stressors such as heat, wind, salt, insects, worms, weevils and weeds.
Erosion resistant	Minimizes soil loss to water and wind.
Drought tolerant	Germinates and produces with minimal water.
Salt tolerant	Thrives and produces in brine water and in soil invaded by irrigation or fertilizer salt.
Ecosystem health – regenerative	
Builds soils	Adds nutrients and organics to the soil rather than constantly extracting.
Low tillage	Requires minimal or no tilling.

Perennial crops	Grows crops that do not have to be replanted every year.

Meaningful breakthroughs in these areas will take decades and cost significantly more than countries are willing to spend. Most these needed innovations for field crops lack even a theoretical model.

Meaningful breakthroughs in these areas will take decades and cost significantly more than countries are willing to spend. Most of these areas lack even a theoretical model. Fortunately, applying new technologies that domesticate microorganisms to assist industrial and organic farmers to enhance soils, nutrient delivery and crop production offer substantial benefits that can be realized now.

Return to organic farming?

Karl Weber, in *Food Inc.: How Industrial Food Is Making Us Sicker, Fatter, and Poorer – and What You Can Do About It* recommends a return to small organic production locally.[242] Unfortunately, the cost structure to sustain a family on a small farm was impossible at least a generation ago and becomes more impractical as crop resources become more expensive. If Wendell Berry cannot support his family on a small farm, it is hard to see how others could succeed.[243] In addition, few Americans are willing to endure the hardship and labor associated with producing food on small farms.

United Nations Under-Secretary-General John Holmes calls for a "new Green Revolution that is agriculturally productive, economically profitable and environmentally sustainable."[244] He says there are no longer any scientific barriers to an immediate and effective response to this urgent clarion call. A systematic, sustained, and successful program of regenerative organic farming needs to follow two guiding principles:

1. Build soil organic matter through the use of cover crops, crop rotation and compost.

2. Improve ecosystem health and human nutrition through plant and animal biodiversity.

Smartcultures

If the transformation to sustainable were simple, far more than 4% of Europeans in 1% of Americans would have changed from industrial farming practices. The acute need is for a pathway that enables modern farmers to transition to eco-friendly practices without disrupting their current farm operations. Smartcultures offers that path.

Smartcultures

Smartcultures are technology neutral in the sense that superior production will require each farmer's thoughtful application of farming methods that are appropriate locally for the weather, crop, soils and markets. While smartcultures enables farmers to transition to predominantly organic methods, some may augment production with selected tools and techniques used in industrial agriculture. Smartculture growers can practice abundant agriculture where the inputs used for growing crops are plentiful and will not run out.

Smartcultures provide substantial value for farmers by saving money, increasing productivity, improving soils and reducing water, energy and fertilizer waste while decreasing erosion and pollution.

Table 10.2 Benefits from Smartcultures

Better plant nutrition and soil structure = better crops

Benefit	Nutrient recovery and delivery
Waste stream	Transform a cost, getting rid of agricultural wastes, to a profit center where solar energy and algae are used in to recover nutrients.
Bioavailable delivery	Deliver nutrients at the right growing cycle stage in bioavailable form that plants can use.
Special delivery	Enable the capacity to deliver specific nutrients such as calcium precisely when they are needed by the crop.

Boost crop quality and productivity	
Texture and taste	Improve produce texture and taste through the immediate bioavailability of nutrients delivered by irrigation.
Productivity	Improve crop yield, speed to maturity, size, weight and quality through nutrients by special delivery.
Vitamins and minerals	Improve the presence, quality and availability of vitamins and minerals in produce by richer nutrient delivery.
Digestible nutrients	Improve the presence of digestible nutrients in produce by using organic biofertilizers.
Upgrade soils	
Soil compaction	Reduce soil compaction and increase prosody to stimulate root growth, make room for microflora and worms to enhance plant strength.
Crust	Strengthen the soil crust to add nutrients, organic material and minimize erosion.
Soil structure	Improve topsoil structure by expanding the humus and organic material in the soil.
Soil microbes	Use algae to attract microbial communities that act to enhance crop health and productivity.

Soil moisture retention	Improve soil moisture retention and decrease heat and drought stress.
Improve agroecology	
Fertilizer pollution	Reduce air, soil and water pollution by using fewer chemical fertilizers.
Erosion	Minimize soil loss to wind and water.
Agricultural chemicals	Minimize pollution from agricultural chemicals and poisons by minimizing or eliminating them.
Bioavailable nutrients	Improve bioavailable nutrients in the soil.
Greenhouse gases	Reduce greenhouse gas emissions, especially CO_2 and nitric oxides.
Tillage	Reduce the need for tillage and soil disruption.
Organic farming	Support and accelerate the transformation of industrial farming to organic farming.
Decrease fossil resource consumption	
Energy	Reduce total energy consumption, especially irrigation, fertilizers and diesel fuel.
Fertilizer	Reduce fertilizer consumption and cost by using reclaimed nutrients.
Water	Reduce water needed for irrigation by improving soil composition and structure.
Irrigation maintenance	Clean calcification from the irrigation system and reduce maintenance costs.

Pesticides, herbicides and fungicides	Reduce the need for pesticides, herbicides, fungicides as well as other agricultural chemicals and poisons.

Nutrient recovery, nutrient delivery and algal seed cultures may be installed now. However, each needs additional R&D to refine the technology and lower costs. The first production units will cost several times as much as later units. A ZooPoo project would focus the global spotlight on smartcultures, especially energy and nutrient recovery, which would benefit everyone on Earth.

Feeding our World

Secretary of State, Hillary Clinton announced a policy shift for the Obama Administration in their commitment to provide leadership in developing a new global approach to hunger. Seven guiding principles will support the creation of effective, sustainable farming systems in regions around the world where current methods are not working:

1. We will seek to increase agricultural productivity, by expanding access to quality seeds, fertilizers, irrigation tools, and the credit to purchase them and training to use them.

2. We will work to stimulate the private sector, by improving the storage and processing of food and improving roads and transportation so small farmers can sell the fruits of their labor at local markets.

3. We are committed to maintaining natural resources, so the land can be farmed well into the future. That includes helping developing communities adapt to climate change, which has had a major effect on the world's farms.

4. We will expand knowledge and training by supporting R&D and cultivating the next generation of plant scientists.

5. We will seek to increase trade so small-scale farmers can sell their crops far and wide.

6. We will support policy reform and good governance, because sustainable agriculture flourishes in a clear and predictable policy and regulatory environment.

7. We will support women and families. 70% of the world's farmers are women, but most programs that offer farmers credit and training target men. This is unfair and impractical. An effective agricultural system must have incentives for those who do the work. And it must take into account the particular needs of those whose futures will shape our world: our children.[245]

These seven principles will guide us and help us set benchmarks to measure the impact of our efforts. Placing sustainable agriculture in national security policy makes sense and has been attempted before under the Carter Administration. Unfortunately, there was little follow-through on prior initiatives. Secretary Clinton's seven principles align with current needs and may succeed if the administration finds a strong sustainable agriculture leader and the announced initiatives are funded.

Production advances in the food supply system, especially those that are regenerative versus extractive, will improve food security. However, meeting the nutritional needs of current and future generations will require far more than additional food production.

Actions that need to be taken include:

Population management. Countries and communities must find ways to slow population expansion. Our planet simply lacks the natural resources for food, water and shelter and transportation for many more people. We are currently over-consuming the Earth's natural resources by over 30%.[246]

Waste management. Roughly half of our food is wasted in the field and along the supply chain. In the US, many consumers waste half of the food they buy. We need to educate farmers, food marketers and consumers and find ways to minimize food, water and energy waste.

Consumer behavior. We need to reverse the trend of higher consumption of resource-expensive meats and dairy products. Available and affordable food will increase as consumers eat more grains, fruits and vegetables.

Recognize and reward food production innovation. Every community, region and nation celebrates their best chefs. Why not do the same for food producers.

Produce locally. Escalating energy costs will make long-distance food transport a distant memory soon. We need to develop local production models that grow food in and near large cities as well as every community. Smartcultures can play a critical role in feeding our children and our global neighbors.

Path forward

Best-practice experiences in permaculture, organic and sustainable agriculture globally drive recommendations for moving forward with smartcultures.

1. End subsidies and protective tariffs for unsustainable and ecological destructive production such as corn ethanol. Subsidies should be shifted to sustainable and ecologically positive production of food, fibers and possibly biofuels.

2. Create a Plant Microbiome Project similar to the National Institute of Health's Human Microbiome Project to identify and genetically sequence the different microbial genera that live in the roots of plants.

3. Develop statutory soil protection policies analogous to those applied to air and water.

4. Recognize and designate algal production as farming. Currently, the USDA does not recognize algal production as farming or aquaculture which means farmers cannot obtain USDA or Farm Credit Services loans.

5. Develop and give public access to a set of metrics from a national smartcultures monitoring system for soil structure, root size and length, water and energy use, crop productivity

improvement, produce quality enhancements, biological diversity, substitution of microbial manufactured fertilizer for chemical fertilizers, avoidance of chemical poisons and pollution mitigation.

6. Additional government commitment may include:
   - Substantial R3D (research, development, demonstration and diffusion).
   - At least one smartcultures demonstration unit in every state.
   - Standards for algal seed culture quality, which may be an extension of existing organic standards.

7. Farmer education may begin with the knowledge transfer from agricultural research universities.
   - Train retired farmers and college students to train farmers.
   - Build a website with training and support materials.
   - Create social networks to support knowledge transfer and best practice.

8. Engage organic gardening, permaculture, botanic gardens, zoos, aquariums, science centers, high schools colleges and universities in smartcultures R3D.

9. Create competitions for the most microbially manufactured N, most efficient nutrient recovery system, best nutrient-delivery model, most efficient irrigation system and other sustainable farming outcomes.

10. Build collaborative R3D with green organizations such as the Sierra Club, the Rodale Institute, World Wildlife Fund, Greenpeace, The Heifer Project, Rotary International and others.

Additional collaborations, ideas, videos and actions are available at the social networking site www.AlgaeCompetition.com.

Summary

Our children's future depends on food security. The two critical threats to food security are global climate change and the mass extinction of fossil resources. Our current food supply will be decimated by the onset of fierce storms, hotter and more acidic oceans, rising sea levels, higher temperatures, hot dry winds and prolonged drought. As the price of fossil fuel rises and the availability of agricultural chemicals diminishes, farmers will need to find sustainable inputs that rely on neither fossil energy nor extracted minerals. Algal biofertilizers offer a potentially low cost and low energy solution for supplying the nutrients needed to grow healthy and hardy crops.

Farmers will embrace the opportunity to transform the costs associated with their waste stream to a profit center with the recovery, recycle and reuse of energy, nutrients and freshwater. Society will celebrate the process of recovering energy and nutrients from waste streams while reversing air, soil and water pollution.

Additional research needs to focus on every smartcultures element. Farmers will also need training in the design, development and operations involved in using algae to remediate their waste streams and deliver nutrients to their crops. New industries will emerge to facilitate both the use of algae to clean and recycle nutrients from waste streams and to deliver those nutrients to crops.

Nature has used algae to support land plants for over 500 million years – since land plants evolved from algae. Now is the time that we need to mimic nature and enable algae to support sustainable, largely organic food production in a manner that uses fewer external inputs, lowers costs, increases productivity and reduces pollution.

Algae are available all over the Earth and are prepared to do their miraculous work to support farmers, hungry consumers and society. We must act quickly to take advantage of algae's green promise – to create food security for everyone on our planet. Together, we can produce sustainable and affordable food while we can leave every field better than we found it.

Appendix I. Soil Macro and Microorganisms[247]

Element	Function in plants
Actino-mycetes	Actinomycetes (ac"-ti-no-my'-cetes) are thread-like bacteria that look like fungi. While not as numerous as bacteria, they perform vital roles in the soil. They help decompose organic matter into humus which slowly releases nutrients. They also produce antibiotics to fight roots diseases. The same antibiotics are used to treat human diseases. Actinomycetes create the sweet, earthy smell of biologically active soil when a field is tilled.
Algae	Thousands of algal species live in the top six inches of soil. Unlike most other soil organisms, algae produce their own food through photosynthesis. However, if there is not enough light, they can use organic nutrients available in the soil. They appear as a green or multicolored film on the soil surface following rain or irrigation. Algae improve soil structure by producing substances that glue soil together into water-stable aggregates. Some species of algae (the blue-greens) can fix their own N, some of which is later released to plant roots.
Arthropods	Arthropods are invertebrate organisms having an external skeleton, a segmented body and jointed appendages. Over a million species have been described in the science literature and they make up more than 80% of all known living species. They range in size from microscopic plankton up to forms a few meters long. Arthropods include the insects, arachnids and crustaceans. Common soil arthropods include sowbugs, millipedes, centipedes, slugs, snails and springtails.

	Arthropods, along with worms, are the primary decomposers. They eat, shred plant and poop animal residues. Some bury residue, allowing other soil organisms to further decompose it. Springtails are a small insect that eat mostly fungi. Their waste is rich in plant nutrients that are released after other fungi and bacteria decompose it. Dung beetles play a valuable role in recycling manure and reducing livestock intestinal parasites and flies.
Bacteria	Most numerous among soil organisms are the bacteria; every gram of soil contains at least a million of these tiny one-celled organisms. There are many different species of bacteria, each with its own role in the soil ecosystem. Bacteria break down complex molecules and enable plants to take up nutrients. Some species release N, S, P and trace elements from organic matter. Others break down soil minerals and release K, P, Mg, Ca and Fe. Other species make and release natural plant growth hormones which stimulate root growth.
	A few bacteria fix N in the roots of legumes while others fix N independently of plant association. Bacteria are responsible for converting N from ammonium to nitrate and back again depending on soil conditions. Various bacteria species increase the solubility of nutrients, improve soil structure, fight root diseases, and detoxify soil.
Earthworms	Earthworm burrows enhance water infiltration and soil aeration. Earthworm tunneling can increase the rate of water entry into the ground 4 to 10 times higher than fields that lack worm tunnels.[248] This reduces water runoff, recharges groundwater, and helps store more soil water for dry spells. Vertical earthworm burrows pipe air deeper into the soil, stimulating microbial nutrient cycling at those

deeper levels. Tillage done by earthworms can replace some expensive tillage work done by machinery.

Worms eat dead plant material left on top of the soil and redistribute the organic matter and nutrients throughout the topsoil layer. Nutrient-rich organic compounds line the tunnels that may remain in place for years if not disturbed. During droughts these tunnels allow for deep plant root penetration into subsoil regions of higher moisture content. In addition to organic matter, worms also consume soil and soil microbes as they move through the soil. The soil clusters they expel from their digestive tracts are known as a worm castings. Each worm cast is separate from other casts and ranges in size from that of a mustard seed to a sorghum seed depending on the size of the worm. The soluble nutrient content of worm casts is considerably higher than those of the original soil (see Table 2). A good population of earthworms can process 20,000 pounds of topsoil per year, with turnover rates as high as 200 tons per acre having been reported in some exceptional cases.

Fungi

Fungi come in many different species, sizes and shapes in soil. Some species appear as thread-like colonies, while others are one-celled yeasts. Slime molds and mushrooms are also fungi. Many fungi aid plants by breaking down organic matter or by releasing nutrients from soil minerals. Fungi are generally early to colonize larger pieces of organic matter and begin the decomposition process. Some fungi produce plant hormones, while others produce antibiotics including penicillin. Several fungi species trap harmful plant-parasitic nematodes.

Mycorrhizae	The mycorrhizae (my-cor-ry-'zee) group of fungi lives either on or in plant roots and act to extend the reach of root hairs into the soil. Mycorrhizae increase the uptake of water and nutrients especially in less fertile soils. Roots colonized by mycorrihizae are less likely to be penetrated by root-feeding nematodes since the pest cannot pierce the thick fungal network. Mycorrhizae also produce hormones and antibiotics, which enhance root growth and provide disease suppression. The fungi benefit from plant association by taking nutrients and carbohydrates from the plant roots where they live.
Nematodes	Nematodes are abundant in most soils and eat decaying plant litter, bacteria, fungi, algae, protozoa and other nematodes and speed the rate of nutrient cycling. A few species are harmful to plants.
Protozoa	Protozoa are free-living microorganisms such as amoeba that crawl or swim in the water between soil particles. Soil protozoa are predatory and feast on other microbes, including bacteria. Protozoa accelerate the cycling of N from the bacteria, making it more available to plants.

Appendix II. The Use of Elements in Plants

Element	Function in plants
	Nonmineral nutrients – source: air and water. Plants use solar energy to change CO_2 and H_2O into chemical energy stored in starches and sugars.
Oxygen, O	• Supports photosynthesis, respiration and plant growth.
Hydrogen, H	• Supplies the H for the creation of hydrocarbons.
Carbon, C	• Supplies the C for the creation of hydrocarbons.
	Macronutrients
Nitrogen, N	Nitrogen plays key roles in all living cells and is a necessary part of all proteins, enzymes and metabolic processes involved in the synthesis and transfer of energy. • Nitrogen is a part of chlorophyll, the green pigment of the plant that is responsible for photosynthesis. • Helps plants with rapid growth, increasing seed and fruit production and improving the quality of leaf and forage crops.
Phosphorus, P	Like nitrogen, phosphorus is an essential part of photosynthesis. • Involved in the formation of all oils, sugars, starches and cellular growth.

	• Helps with the transformation of solar energy into chemical energy, plant growth and maturation and helps plant withstand stress. • Encourages rapid growth, blooming, fruiting and root growth. • Drives ATP, DNA and phospholipids.
Potassium, K	Potassium is absorbed by plants in larger amounts than any other mineral element except nitrogen and, in some cases, calcium. • Helps in building plant proteins. • Essential for photosynthesis, fruit quality and protection from disease.
Magnesium, Mg	Magnesium is part of the chlorophyll in all green plants and essential for photosynthesis. • Helps activate many plant enzymes needed for growth. • Oxygen evolving complex of photosystem II • Aids in cell wall formation
Sulfur, S	• Essential plant food for production of plant proteins. • Promotes activity and development of enzymes and vitamins. • Helps in chlorophyll formation. • Improves root growth and seed production. • Helps with vigorous plant growth and resistance to cold.

	Micronutrients
Boron, B	• Unique among micronutrients because plants require it in a narrow range of concentrations in the rhizosphere . • Helps in the use of nutrients and regulates other nutrients. • Aids in production of sugar and carbohydrates. • Essential for seed and fruit development. • Critical for cell wall and root tip growth.
Copper, Cu	• Important for reproductive growth. • Aids in root metabolism and helps in the utilization of proteins. • Acts as a cofactor for several enzymes and lastocyanin, cytochrome oxidase.
Chlorine, Cl	• Aids plant metabolism in photosynthesis. • Oxygen production in photosynthesis. • Regulates transpiration.
Iron, Fe	• Essential for the formation of chlorophyll. • Acts as a cofactor for several enzymes. • Serves as a base in many plant molecules.
Silicon, Si	• Nitrate reductase • Supports cell wall formation.
Sodium, Na	• Chlorophyll functions
Magnesium, Mg	• Operates with enzyme systems involved in

	the breakdown of carbohydrates and nitrogen metabolism.
Molybdenum, Mo	• Necessary for nitrate reduction. • Controls symbiotic nitrogen fixation.
Zinc, Zn	• Essential for the transformation of carbohydrates. • Regulates consumption of sugars. • Key part of the enzyme and hormone systems that regulate plant growth. • Cofactor with enzymes and regulates protein synthesis. • Critical for plant reproduction.
Cobalt, Co	• Aids in building Vitamin B_{12}.
Vanadium, V	• Bromoperoxidase and some nitrogenases.
Bromine, Br	• Halogenated compounds
Iodine, I	• Protects plant with antimicrobial, anti-herbavore or allelopathic functions

Acknowledgements

Thanks first to my best friend and life partner, Ann Ewen, who made this project possible. Ann supported both the "ah's" and the "aha!'s"

This product would not have been possible without the extraordinary research of David and Marcia Pimentel, Lester Brown, President of the Earth Policy Institute, Jeffery Sachs of the Earth Institute. Professors Qiang Hu, Milton Sommerfeld and Bruce Rittman supported endless questions on molecular biology and algae production. Environmental scientists Al Darzins, Eric Jarvis and Mike Siebert at NREL were very helpful with renewable energy sources. Thanks also to great advisors who elevated *smartcultures* from a solo to an orchestra. Kate Smith and Darla Winfrey helped with proofing.

Science	Business – Econ.	Agribusiness
• Bob Thompson	• Robert Henrikson	• Jon Ewen
• Dan Childers	• Mark Allen	• Gary Wood
• James Elser	• Alan Resnik	• Doug Young
• Jessica Corman	• Susan Schultz	• Jim Robertson
• Herb Roskind	• Gordon LeBlanc, Jr.	• Tracy Penwell
• Andy Ayers	• William Cockayne	• Barry Spiker

Also helpful were the published works of Paul Ehrlich, Sandra Postel, Nobel Laureate Al Gore, Harvey Blatt, Fred Pearce, Michael Pollen and Linda Graham. High-content websites were a great support such as Algaebase, U.N., W.H.O., the National Resources Defense Council, Sierra Club, Green Peace, Audubon Society, Union of Concerned Scientists, Center for Energy and Climate Solutions, Clean Water Network and Public Citizen. Also useful were U.S. government sources including: DOE, EPA, USDA, NOAA and NREL.

Mark Edwards

Mark works on sustainable and affordable food and energy (SAFE) production that is available to everyone on Earth. Mark's goal is to improve food production and help growers leave every field better than they found it.

Hunger, nutrition, pollution and soil regeneration are urgent global challenges. Abundant agriculture and the use of smartcultures provide a practical path for improving our nutrition, lowering food costs and ending ecological pollution. We need to engage millions of Green Masterminds globally who have the capacity to grow algae as food, feed and fertilizer for the needs of their family and community.

Mark graduated from the US Naval Academy in mechanical engineering, oceanography and meteorology where Jacques Cousteau motivated and mentored his interest in the oceans and global stewardship. He holds an MBA and PhD in marketing and consumer behavior and has taught agribusiness food marketing, sustainability and entrepreneurship at Arizona State University for over 30 years.

Mark served as CEO of TEAMS Intl. for 24 years, the software and assessment firm he founded based on his research on advanced assessment technologies, talent and leadership assessment. He served as lead consultant for more than 400 firms globally. He was retained by many US departments and the military, including DOE, DOD, Special Forces and the National Labs.

Mark served as a Director for a Fortune 50 foods company and has done extensive R&D on new foods, sources and consumer behavior. He has consulted for Monsanto, Pioneer Seeds, DuPont, Nabisco, Quaker Oats, General Mills, Borden and many other agribusiness companies. He has worked with senior executives at 15 large US oil and gas firms as well as British Petroleum and Saudi Aramco.

The Green Algae Strategy Series
By Mark R. Edwards

THE GREEN ALGAE STRATEGY SERIES focuses on creating Sustainable and Affordable Food and Energy – "SAFE" production. THE GREEN ALGAE STRATEGY SERIES are available for free downloading for teachers, professors and students in color PDF at www.AlgaeAlliance.com. They are also available on Amazon.com.

BioWar I: Why Battles Over Food and Fuel Lead to World Hunger. 2007. BioWar I, where food is burned for fuel, must be ended by withdrawal – not of soldiers, but of ecologically damaging agricultural subsidies. The unintended consequences on U.S. and world food markets of subsidies that promote the production of ethanol from corn will be catastrophic for the U.S. as we run out of fossil water, soils, air, and food exports, and food prices escalate.

Green Algae Strategy: End Oil Imports and Engineer Sustainable Food and Fuel. 2008. Algae offer solutions for sustainable and affordable food and energy because algae are the most productive biomass source on Earth. Fossil-based agriculture is non-sustainable because it uses far too many nonrenewable resources. Algae offer sustainable food, fuel, fertilizer, fine medicines and a pollution solution. Named *Best Science Book – 2009, Independent Publishers.*

Green Solar Gardens: Algae's Promise to End Hunger. 2008. Algaculture in small solar gardens distributed globally will enable SAFE production locally. *Green Solar Gardens* addresses the web of poverty and hunger, including affordable food, fodder, fish food, fertilizer, fire for cooking and heating and fine medicines.

Crash! The Demise of Fossil Foods and the Rise of Abundance. 2009. Traditional fossil-based agriculture sits precariously on a foundation of unsustainable fossil resources that will become unaffordable and then will run out. Abundant agriculture is sustainable because it uses plentiful inputs that are cheap, will stay affordable and available.

Great Green Reading

Food, energy and economics

Thomas L. Friedman, *Hot, Flat, and Crowded: Why We Need a Green Revolution – and How It Can Renew America*, Farrar and Giroux, 2008.

Lester R. Brown, *Plan B 3.0: Mobilizing to Save Civilization*, Third Ed., W. W. Norton; 2008.

Jeffrey D. Sachs, *Common Wealth: Economics for a Crowded Planet*, Penguin Press HC, 2008.

Jeffrey D. Sachs, *The End of Poverty: Economic Possibilities for Our Time*, Penguin Press, 2005.

Fred Krupp and Miriam Horn, *Earth: The Sequel: The Race to Reinvent Energy and Stop Global Warming*, W. W. Norton, 2008.

Brangien Davis and K. Wroth, *Wake Up and Smell the Planet: Non-Preachy Grist Guide to Greening Your Day*, Mountaineers Books, 2007.

Daniel Esty and Andrew Winston, *Green to Gold: How Smart Companies Use Environmental Strategy*, Yale University Press, 2006.

Water

Elizabeth Kolbert, *Field Notes from a Catastrophe: Man, Nature, and Climate Change*, Bloomsbury, 2006.

Sandra Postel, *Pillar of Sand: Can the Irrigation Miracle Last?* W. W. Norton & Company, 1999.

Peter H. Gleick, *The World's Water 2006-2007*: The Biennial Report on Freshwater Resources, Island Press, 2006.

Fred Pearce, *When the Rivers Run Dry: Water –The Defining Crisis of the Twenty-first Century*, Beacon Press, 2007.

Vandana Shiva, *Water Wars: Privatization, Pollution, and Profit*, South End Press, 2002. Robert Jerome Glennon, *Water Follies: Groundwater Pumping and The Fate Of America's Fresh Waters*, Island Press, 2004.

Index

Abundant agriculture, 62, 64, 193
acidic seawater, 78
Africa, 99
agricultural chemicals, x, xi, xii, 4, 5, 7, 14, 15, 16, 17, 18, 19, 21, 38, 39, 40, 42, 48, 52, 53, 83, 90, 91, 93, 95, 102, 122, 127, 157, 171, 176, 177, 181
Al Darzins, 191
algaculture, 141
Algal Biomass Organization, 145
algal production, 141
algalization, 17, 161
antibiotics, 44, 51, 183, 185, 186
antioxidants, 11, 12, 127, 168
aquaculture, 62
arthropods, 22, 183
Ashworth, 166
Australia, 77, 79
Azolla, 159

BASF, 169
beta-carotene, 11
bioavailable nutrients, 3, 4, 6, 8, 22, 29, 176
biodiversity, xii, 22, 44, 46, 48, 78, 86, 87, 95, 105, 173
biofertilizer, xiii, 17, 18, 54, 67, 123, 135, 147, 158, 159, 162, 163, 169, 175, 181
biofuel, 64, 141, 166
Biological pest control, 5
biotechnology, 139
Biowar I, 193
blue-green algae, 17, 67, 69, 70, 104, 157, 160, 161

CAPS, 124, 131, 132, 133, 135, 136, 137, 138, 140, 141, 142, 144
Center for Biological Diversity, 43
chemical fertilizers, xii, 5, 7, 11, 12, 13, 15, 17, 18, 24, 41, 42, 55, 67, 104, 149, 151, 153, 154, 162, 163, 171, 176, 180
China, xiv, 17, 41, 77, 82, 90, 93, 99, 109, 129, 157
Clinton, Hillary 177
Community starvation, 90

conservation agriculture, 60
cotton, 12, 155, 160
cover crops, 45, 48, 50
crop rotation, 5, 57, 97, 106
cyanobacteria, 141

dead zone, 42, 105
Death by Supermarket, 10
Diamond, Jared, 36
dilution effect, 11
DNA, 107, 171, 187
DOE, 145, 191, 192
Drip irrigation, 54

ecosystem, ix, 1, 2, 6, 42, 46, 53, 61, 66, 78, 117, 131, 163
Elizabeth Kolbert, 36
Empty calorie foods, 10
Ending Hunger Now, 165
erosion, 4, 5, 9, 11, 12, 13, 15, 16, 24, 25, 26, 27, 29, 30, 35, 36, 43, 45, 48, 54,
    56, 60, 68, 72, 78, 79, 90, 95, 96, 97, 106, 147, 149, 152, 166, 168, 174, 175
ethanol, 193
European Union, 48, 50, 169

fertile soil, 22, 95
fertilizer extinction, 114
food cascade, 84, 114
Food Inc., 173
food security, 169
food supply, x, xi, xii, 2, 3, 9, 19, 35, 43, 44, 58, 74, 93, 103, 114, 117, 178, 181
fossil energy, 16, 40, 91, 129, 151, 171, 181
fossil nutrient problem, 106
fossil resources, ix, xi, xii, xiv, 2, 3, 14, 37, 39, 40, 62, 74, 87, 89, 90, 91, 94, 102,
    110, 111, 112, 116, 126, 132, 135, 157, 165, 166, 167, 181, 193

genetically engineered seeds, 17, 105, 170
genetically modified crops, 38, 45, 93, 102, 126, 133, 169
global warming, 73, 86, 117, 126, 127
Green Algae Strategy, iv, 193
Green Independence, 192
Green Revolution, 37, 45, 93, 173, 194

greenhouse gases, 13, 105
health, 4, 9, 10, 11, 15, 16, 21, 22, 44, 46, 48, 59, 82, 87, 91, 95, 104, 105, 118,
    122, 126, 131, 150, 151, 154, 166, 170, 171, 172, 173, 175
heat stress, 75
herbicides, 43, 87, 105
Human Microbiome Project, 21, 179
humus loss, 29, 96
hydroponic, 31, 61, 62

India, 17, 19, 28, 77, 82, 83, 93, 94, 99, 109, 157, 158, 159, 160, 161, 162, 167
industrial agriculture, 2, 9, 10, 14, 25, 39, 41, 44, 86, 95
Iowa, 105

Kolbert, Elizabeth, 36
krill, 78

Mexico, 99
microalgae, 3, 28, 30, 31, 62, 154, 157
microbial communities, 2, 15, 21, 23, 41, 175
microfarms, 64
micronutrient deficiencies, 9
micronutrients, 3, 6, 8, 10, 11, 12, 13, 14, 16, 23, 40, 54, 56, 57, 68, 69, 70, 71,
    91, 125, 127, 133, 150, 153, 154, 157, 168, 172, 189
microorganisms, vii, xii, xiii, 6, 22, 23, 24, 25, 27, 36, 38, 41, 63, 67, 73, 143, 44,
    148, 149, 152, 157, 158, 159, 160, 162, 168, 173, 186
Millennium Project, 168
moisture retention, 28, 48, 55, 68, 71, 95, 150, 176
monocropping, 86
monocultures, xii, 43, 127
Monsanto, 74, 169, 170, 171, 192
mycorrhizae, 161

N fertilizer, 38, 40, 42, 91, 102, 104, 159
NASA, 62, 73
natural processes, xiii, 17, 31, 95, 103, 169
Net zero exports, 101
No-till farming, 5, 97
N-P-K fertilizers, 12, 40
NREL, 191
nutrient delivery, xiii, 17, 18, 19, 56, 57, 58, 132, 140, 150, 173, 175, 177

nutrient erosion, 3, 166
nutrient recovery, 3, 17, 19, 114, 115, 118, 135, 136, 140, 177, 180

Ogallala Blue, 166
organic agriculture, xiii, 46, 54
organic capital, 4, 5, 67
organic farming, 3, 17, 19, 29, 45, 46, 47, 48, 50, 52, 67, 163, 165, 173, 176
Organic fertilizers, 50, 147, 149, 151
organic matter, 11, 12, 25, 26, 27, 28, 29, 41, 45, 47, 48, 50, 52, 53, 54, 65, 69,
    95, 96, 97, 104, 121, 149, 151, 160, 161, 163, 173, 183, 184, 185
organic production, 7, 16, 35, 50, 52, 53, 60, 62, 125, 173

permaculture, 61, 179, 180
pesticides, 2, 5, 12, 15, 27, 38, 39, 43, 44, 46, 47, 48, 55, 59, 60, 83, 87, 90, 93,
    102, 105, 106, 127, 147, 150, 166, 177
pH, 142
phosphate rock, 42, 109, 110
phosphorus, 9, 103, 107, 187
photosynthesis, 140, 189
phytochemicals, 10, 11, 12, 16
Pimentel, David, 40
plant growth hormones, 3, 136, 152, 153, 184
plant growth regulators, 23, 31, 34, 151
plant resistance, 44
polysaccharide sheaths, 29, 67, 69, 71
profit center, 4, 117, 127, 131, 174, 181
pyrolysis, 4, 119, 121, 122, 123, 128

regenerative, vii, x, xii, 48, 172, 173, 178
resource extinction, ix, 93, 130
resource run, 84, 111, 112, 114
rhizobia bacteria, 70, 159
Romans, 69
roots, vii, xii, xiii, 6, 10, 11, 12, 22, 23, 24, 32, 38, 41, 42, 45, 54, 61, 69, 70, 71,
    72, 104, 133, 148, 152, 158, 159, 160, 161, 179, 183, 184, 186

Sachs, Jeffrey, 167
salt, 9, 10, 27, 41, 57, 74, 77, 79, 90, 97, 98, 117, 126, 139, 150, 169, 172
San Joaquin Valley, 99
seaweed, 31, 32, 62, 69, 153

social stability, 81
soil chemotherapy, 7
soil conditioners, v, 27, 136, 153
soil fertility, ix, 5, 22, 25, 35, 46, 47, 50, 54, 60, 68, 108, 148, 157, 159, 162
soil nutrients, xi, xii, 5, 7, 11, 12, 14, 26, 35, 45, 48, 56, 69, 138
soil organics, 2, 12, 14, 17, 25, 29, 35, 36, 48, 55, 56, 57, 66, 70, 106, 150
South Africa, 99
soybean disease, 86
strain selection, 139
subsidies, 193
supply disruption, 84, 111, 112
sustainable agriculture, 59
sustainable food, 193
symbiotic relationship, 70

Texas, 139
The End of Poverty, 167
tomato blight, 74
Tomorrow's Table, 53
transgenic seeds, 169

U.C. Davis, 47
United Nations Environmental Program, 46
USDA, 191

Venezuela, 101

Washington State, 30, 51
water pollution, 5, 44, 51, 148, 168, 176, 181
wetlands, 140
World Bank, 46, 74

ZooPoo360, 116

---

[1] Edwards, Mark R. *BioWar I: Why Battles over Food and Fuel Lead to World Hunger*, Tempe: LuLu Press, 2008.

[2] CNN, Riots, instability spread as food prices skyrocket, June 10, 2008. http://www.cnn.com/2008/WORLD/americas/04/14/world.food.crisis/

[3] WRI, World Resources Institute. New York: Oxford University Press, 1994.

[4] There is no clear fossil record but today, algae crusts are common on deserts and sandy beaches. Algae's cousins, mosses and liverworts support the algal bridge theory because they have no roots, yet are able to elongate cells to provide some root functions.

[5] Shapouri, Hosein, et al, The 2001 net energy balance of corn ethanol. US Department of Agriculture: office of chief economist, agricultural research service, Washington, DC, 2004.

[6] Khan, Mohammad Saghir, et. al. Plant growth promotion by phosphate solubilizing fungi - current perspective, *Archives of Agronomy and Soil Science,* 56: I, February 2010, 73-98

[7] Lawrence, F. "214". in Kate Barker. Not on the Label, Penguin, 2004, 213.

[8] FAO – malnutrition.

[9] Roberts, Paul. The End of Food by Paul Roberts, 2009.

[10] Halweil, Brian. Still No Free Lunch: Nutrient levels in U.S. food supply eroded by pursuit of high yields, Organic Center, September 2007.

[11] Deville, Nancy. Death by Supermarket: The Fattening, Dumbing Down, and Poisoning of America, Barricade Books, 2007.

[12] Bokhtiar, S.M. & Sakurai, K. 2005. Effects of organic manure and chemical fertilizer on soil fertility and productivity of plant and ratoon crops of sugarcane. Archives of Agronomy and Soil Science, 51: 325-334.

[13] Mitchell, C.C., and J.A. Entry. 1998. Soil C, N and crop yields in Alabama's long-term 'Old Rotation' cotton experiment.Soil Tillage Res. 47:331–338.

[14] Dias De Oliveira, Marcelo E., Burton E. Vaughan, and Edward J. Rykiel Jr. "Ethanol as Fuel: Energy, Carbon Dioxide Balances, and Ecological Footprint." BioScience 55:7 (2005): 593–602.

[15] India's agal experience.

[16] Harman, Katherine. Bugs Inside: What Happens When the Microbes That Keep Us Healthy Disappear? Scientific American, December 16, 2009.

[17] Magdoff, Fred and Ray R. Weil. Soil Organic Matter in Sustainable Agriculture (Advances in Agroecology), New York: CRC Press 2004, 2.

[18] Tisdale, S. L. and W. L. Nelson. *Soil Fertility and Fertilizers*. 3rd ed. New York: Macmillan, 1975.

[19] Brady, N. C. *The Nature and Properties of Soils*. New York: Macmillan Publishing Co., 1974.

[20] Gliessman, Stephen R. Agroecology: The Ecology of Sustainable Food Systems, Second Edition, CRC Press; 2 ed., 2006.

[21] Plaster, E. J. *Soil Science and Management*. 3rd ed. Albany: Delmar Publishers, 1996.

[22] Magdoff, Fred and Ray R. Weil. Soil Organic Matter in Sustainable Agriculture (Advances in Agroecology), New York: CRC Press, 2004, 25.

[23] Coleman, D.C. and D.A. Crorsley, Jr., 2004,Fundamentals of Soil Ecology, 2nd edition, New York: Academic Press.

[24] Barber, S. A. *Soil Nutrient Bioavailability: A Mechanistic Approach*. New York: Wiley, 1984.

[25] Hempfling, R., Schulten, H.R., Horn, R., Relevance of humus composition to the physical/mechanical stability of agricultural soils. Journal of Analytical and Applied Pyrolysis 17:, 1990,.275–281.

[26] Rees, R.M. B. C. Ball, C. D. Campbell, and C. A. Watson. Sustainable Management of Soil Organic Matter, CABI, 2001.

[27] Sullivan, Preston. Appropriate Technology Transfer for Rural Areas, (ATTRA). http://www.soilandhealth.org/01aglibrary/010117attrasoilmanual/010117attra.html#2

[28] Daudu, Christogonu.s Organic Matter Sources, Soil Fertility and Productivity, VDM Verlag, 2008.

[29] Ibid.

[30] Anon. Product choices help add to worm counts. Farm Industry News. February, 1997, 64.

[31] Allison, F.E. soil organic matter and its role in crop production. New York: Elsevier, 1973.

[32] Scherr, Sara and Sajal Sthapit. Farming and land use to cool the planet, in The State of the World 2009, the World Watch Institute, 2009, 37.

[33] Kolbert, Elizabeth. Field Notes from a Catastrophe: Man Nature and Climate Change. New York: Bloomsbury, 2006: 96-97.

[34] Diamond, Jared. Collapse: How Societies Choose to Fail or Succeed. New York: Viking Adult, 2004, 54.

[35] Personal interview, Marvin Morrison, July, 2003.

[36] Pimentel, David and Marcia Pimentel. *Food, energy and society*, 3rd Ed., Ney York: CRC Press, 2008, 43.

[37] Pimentel, D. and Patzek, T. Natural resources research, 14;1, 65 – 76, 2005.

[38] Francis, Charles, A., Cornelia B. Flora, and Larry D. King. Sustainable Agriculture in Temperate Zones. John Wiley and Sons, Inc. New York. 1990, 487.

[39] Parker, M.B., G.J. Gasho, and T.P. Gaines. Chloride toxicity of soybeans grown on Atlantic coast flatwoods soils. Agronomy Journal. 75, 1983, 439-443.

[40] USDA ERS, Fertilizer Use, 2008. http://www.ers.usda.gov/Data/FertilizerUse/

[41] Sawyer, John. Natural gas prices affect nitrogen fertilizer costs, 2001.http://www.ipm.iastate.edu/ipm/icm/2001/1-29-2001/natgasfert.html

[42] Earth Policy Institute, Earth Policy Institute, Oil and Food: A Rising Security Challenge, Earth Policy Institute, 2005.

[43] Environmental New Service, EPA Faces Lawsuits, 2 Feb 2010. http://www.ens-newswire.com/ens/feb2010/2010-02-02-091.html

[44] Raloff, Janet. EPA reviews hint of weed killer's fetal risks, *Science News*, February 4th, 2010.

[45] Editors, Weed killer feminizes fish, *Science News*, Nov 4, 2002. 275.

[46] Halweil, B. "Pesticide-Resistance Species Flourish." Vital Signs. Ed. L. Starke. New York: Norton, 1999: 124.

[47] "In Brief." Environment, Sept. 2001: 8.

[48] Shah, Sonya. Behind Mass Die -Offs, Pesticides Lurk as Culprit, Yale360, Jan 7, 2010. http://www.e360.yale.edu/content/feature.msp?id=2228

[49] Cimitile, Matthew, Crops absorb livestock antibiotics, science shows, Environmental Health News, January 6, 2009.

[50] Fossel, Peter V. Organic Farming: Everything You Need to Know, Voyageur Press, 2007.

[51] Pimentel, David. Food, energy and society, 29.

[52] USDA Study Team on organic farming, 98 – 99, cited in USDA ARS. Organic Farming Beats No-Till?, 2007.

[53] United Nations Environmental Programme (UNEP) and United Nations Conference on Trade and Development (UNCTAD), Organic Agriculture and Food Security in Africa, 2008, 1-61.

[54] IAASTD, 2008. International Assessment of Agricultural Knowledge, Science and Technology for Development Global Report.

[55] Pretty, J., A. D. Noble, D. Bossio, J. Dixon, R. E. Hine, F. W. T. Penning de Vries, and J. I. L. Morison .2006. Resource-

Conserving Agriculture Increases Yields in Developing Countries. Environmental Science and Technology, 40: 4.

1114-1119. www.ecofair-tradeorg/pics/en/brosch_ecofairtrade_el.pdf

[56] Maeder, P. *et al.* (2002). *Soil Fertility and Biodiversity in Organic Farming.* Science v296, , 1694-1697.

[57] Woese, K. et al. A comparison of organically and conventionally grown foods -- results of her review of the relevant literature. Journal of science in food and agriculture, 74:3, 1997, 281 -- 293.

[58] Lampkin, N. H. estimating the impact of widespread conversion to organic farming on land use and physical output in the United Kingdom. In: The economics of organic farming, N.H. Lampkin and S. Padel, Eds, CAB: Wallingford, UK, 1994, 343 -- 359.

[59] Clark, S., Klonsky, K., Livingston, P., Temple, S. 1999. Crop-yield and economic comparisons of organic, low-input, and conventional farming systems in California's Sacramento Valley. American Journal of Alternative Agriculture 14: 109-121.

[60] USDA ARS. Organic Farming Beats No-Till?, 2007.

[61] Kirchmann H et al. (2007). Comparison of Long-Term Organic and Conventional Crop-Livestock Systems on a Previously Nutrient-Depleted Soil in Sweden. *Agronomy Journal* 99:960-972

[62] organics, Tim Lobo Rodale

[63] LaSalle,Tim, Paul Hepperly and Amadou Diop The Organic Green Revolution, Rodale Institute, 2008, 5.

[64] Liebig, M., Doran, J. 1999. Impact of organic production practices on soil quality indicators. J. Environ. Quality

28(5):1601–1609. Lotter, D., Seidel, R., and W. Liebhardt. 2003. The performance of organic and conventional cropping systems in an extreme climate year. American Journal of Alternative Agriculture 18(2):1-9.

[65] Veenstra, J., Horwath, W., Mitchell, J., Munk, D. 2006. Conservation tillage and cover cropping influences soil properties in San Joaquin Valley cotton-tomato crop. California Agriculture 60(3):146-152.

[66] Hanson, J., Lichtenberg, E., and S. Peters. 1997. Organic versus conventional grain production in the mid-Atlantic: An economic overview

and farming system overview. American Journal of Alternative Agriculture 12(1):2-9.

[67] Azeez, G. 2008. The comparative energy efficiency of organic farming. Colloquium on Biologically Based Agriculture and the Climate. Enita Clermont France April 17-18, 2008. 7 pages.

[68] Teasdale, J., Coffman, C., Mangum, R. 2007. Potential long-term benefits of no-tillage and organic cropping systems for grain production and soil improvement. Agronomy Journal 99, 2007.

[69] Reider, C., Herdman, W., Drinkwater, L., and R. Janke. 2000. Yields and nutrient budgets under composts, raw dairy manure and mineral fertilizer. Compost Science and Utilization 8(4):328-339.

[70] Pimentel D, Hepperly P, Hanson J, Douds D, Seidel R. Environmental, energetic, and economic comparisons of organic and conventional farming systems. BioScience 2005, 55: 573-582.

[71] Douds, D., Hepperly, P., Seidel, R., and K. Nichols. 2007. Exploring the role of mycorrhizal fungi in carbon sequestration in agricultural soil. American Society of Agronomy Annual Meeting, November, New Orleans, Abstr. 1011. Buyer, J., and Kaufman, D. 1997. Microbial diversity in the rhizosphere of corn grown under conventional and low-input systems Applied Soil Ecology 5(1): 21-27.

[72] Delate, K., Turnbull, R., and DeWitt, J. 2006. Measuring and Communicating the Benefits of Organic Foods. Crop Management doi: 10.1094/CM-2006-0921-14-PS. Delate, K., Turnbull, R., and DeWitt, J. 2006. Measuring and Communicating the Benefits of Organic Foods. Crop Management doi: 10.1094/CM-2006-0921-14-PS.

[73] Mitchell, A., Hong, Y, Koh, E., Barrett, D., Byrant, D., Denison, R., and S. Kaffka. 2007. Ten year comparison of influence of organic and conventional crop management on contents of flavonoids in tomatoes. J. Agric. Food Chem. 55(15):6154-6159.

[74] Lu, C., Toepel, K., Irish, R., Fesnke, R., Barr, D., and R. Bravo. 2005. Organic diets significantly lower children dietary exposure to organophosphorus pesticide. Environ. Health Perspect. 114(2): 260-263. Curl, C., Fenske, R., and K. Elgethun. 2003. Organophosphorus pesticide exposure of urban

and suburban pre-school children with organic and conventional diets. Environ. Health Perspect. 111:377-382.

[75] http://www.organic-world.net/

[76] Willer, H. and Klicher, L. (Eds.), (2009): The World of Organic Agriculture. Statistics and Emerging Trends 2009. IfOM, Bonn, FiBL, Frick, ITC, Geneva, www.organic-world.net

[77] Van Horn, Mark. Compost production and utilization: a grower's guide. UCDANR publication, 21514, 1995.

[78] Gillman J. *The Truth about Organic Farming*. Timber Press. 2008.

[79] Firmani, Mark. Chemical Companies Slapped with Multimillion Dollar Class Actions for Poisoning Fields, www.ReglanLawsuitsCenter.com.

[80] Associated Press, Tyson Foods included in water suit, *Texarkana Gazette*, 01/02/2009.

[81] Union of Concerned Scientists, Industrial Agriculture: Features and Policy, 2009.

[82] Cimitile, Matthew. Crops absorb livestock antibiotics, science shows. Environmental Health News, January 6, 2009.

[83] Westerman, P.W., Safley Jr., L.M., Barker, J.C., 1983. Potential harmful effects from land applying animal manures. ASAE Paper No. 83-2612 presented at ASAE Meeting, Chicago, IL, ASAE, St. Joseph, MI.

[84] Nebraska company poised to lose organic certification. New York Times, December 2, 2009.

[85] Ronald, Pamela and Raoul Adamchak. *Tomorrow's Table: Organic Farming, Genetics and the Future of Food*,

[86] National Sustainable Agriculture Coalition (NSAC). http://sustainableagriculture.net/

[87] Brundland, G.H. Our Common Future. Report of the world commission on the environment and development. Oxford: Oxford University press, 1987.

[88] Horrigan, Leo, Robert Lawrence and Polly Walker. How sustainable agriculture can address the environmental and human health harms of industrial agriculture, Env. Health Perspectives, 110:5, May, 2002.

[89] U.C. Sustainable Agriculture Research and Education Program, 1997. http://sarep.ucdavis.edu/concept.htm

[90] Ikerd, John. *Crisis and Opportunity: Sustainability in American Agriculture,* University of Nebraska Press, 2008, 71.

[91] Ken E. Giller, Ernst Witter, Marc Corbeels, Pablo Tittonell . Conservation agriculture and smallholder farming in Africa: The heretics' view. Field Crops Research, Volume 114, Number 1 (October 2009), pp. 23-34.

[92] Ibid.

[93] Bell, Graham. *The Permaculture Way*, 2st edition, Permanent Publications, 2004.

[94] Hoagland, Dennis R. and Daniel I. Arnon. The Water Culture Method for Growing Plants Without Soil, Agricultural bulletin, 1938.

[95] Stoner, R.J. and J.M. Clawson (1997-1998). *A High Performance, Gravity Insensitive, Enclosed Aeroponic System for Food Production in Space.*Principal Investigator, NASA SBIR NAS10-98030.

[96] Edwards, Mark R. *Abundance: Our Future Fossil-free, Clean and Sustainable Food and Energy,* Tempe: CreateSpace, 2011.

[97] Edwards, Mark R. Green Algae Strategy: End Oil Imports and Engineer Sustainable Food and Fuel, CreateSpace, 2009, 98.

[98] Edwards, Mark R. Crash! The Demise of fossil Foods and the Rise of Abundance, CreateSpace, 2009, 283.

[99] Edwards, Mark R. Green Solar Gardens: Algae's Promise to end Hunger, CreateSpace, 2009, 283.

[100] Goyal, SK. A profile of algal biofertilizer. In in Kannaiyan, S. Ed., Delhi: Narosa Publishing House, 2002, 250 – 258.

[101] Light penetration in soils.

[102] Kuepper, George. Foliar Fertilization. NCAT Agriculture Specialist, 2003. http://attra.ncat.org/who.html

[103] Kannaiyan, Introduction to Biofertilizers, in Biotechnology of biofertilizers, in Kannaiyan, S. Ed., Delhi: Narosa Publishing House, 2002, 4.

[104] Harvel, C.D., et al. Climate warming and disease risk for terrestrial and marine biota. Science, 296:2158–2162.

[105] Intergovernmental Panel on Climate Change (IPCC). Climate change 2007: the physical science basis, ed R. B. Alley, T. Bernsten, N.L. Bindoff, et al.

[106] James Hansen et al., "Climate Change and Trace Gases," Philosophical Transactions of the Royal Society A, vol. 365 (2007), pp. 1925–54.

[107] http://www.cnn.com/2008/WORLD/americas/04/14/world.food.crisis/

[108] Zoellick, Robert. High-Level Conference on World Food Security, Rome, World Bank, June 4, 2008. No: 2008/349/EXC.

[109] http://www.triplepundit.com/pages/un-rome-conference-mobilisatio-003225.php

[110] Today in Biofuels, Biofuels Digest, http://www.biofuelsdigest.com/blog2/2008/03/07/today-in-biofuels

[111] http://www.larouchepac.com/node/10736

[112] http://deltafarmpress.com/news/food-procudtion-0604/

[113] Annual Energy Report" (PDF). US Department of Energy (2006-07).

[114] World Bank, World Development Report 2008: Agriculture for Development, October, 2007. http://publications.worldbank.org/ecommerce/catalog/

[115] Nicholas Stern, The Stern Review on the Economics of Climate Change (London: HM Treasury, 2006).

[116] Mean temps since 1970.

[117] Moskin, Jullia. Outbreak of Fungus Threatens Tomato Crop, *New York Times*, July 17, 2009.

[118] Mohan K. Wali et al., "Assessing Terrestrial Ecosystem Sustainability," Nature & Resources, October–December 1999, pp. 21–33.

[119] Lobel, David and Gregory Asner. "Climate and Management Contribution to Recent Trends in US Agricultural Yields." Science 299 (14 Feb. 2003): 1032.

[120] J. Larsen, Earth Policy Institute, published online 28 July 2006 www.earth-policy.org/Updates/2006/Update56.htm

[121] W. Easterling et al., in Climate Change 2007: Impacts, Adaptation, and Vulnerability, M. Parry et al., Eds. (Cambridge Univ. Press, New York, 2007), p. 976.

[122] Roosevelt, Margot. Climate change could put the heat on California crops, LA Times, July 22, 2009.

[123] Tankersley, Jim. California farms, vineyards in peril from warming, U.S. energy secretary warns, Los Angeles Times, February 4, 2009.

[124] Scherr, Sara and Sajal Sthapit. Farming and land use to cool the planet, in The State of the World 2009, the World Watch Institute, 2009, 31.

[125] World Bank, World Development Report 2008:

Agriculture for Development, World Bank, Washington, DC, 2007.

[126] Battisti, David and Rosamond Naylor. Historical Warnings of Future Food Insecurity with Seasonal Heat, Science, 8 September 2008,10.1126

[127] Pennisi,Elizabeth. Western U.S. Forests Suffer Death by Degrees, Science 23 January 2009: 447.

[128] Cave, Damien. Cold Still Hurting Florida's Farmers, New York Times, January 13, 2010.

[129] Food and Agricultural Organization, FAO. Dire winter triggers livestock disaster in Mongolia. http://www.fao.org/

[130] Knutson, Thomas R., John L. McBride, et. al. Tropical cyclones and climate change, Nature Geoscience, 21 February 2010, doi:10.1038/ngeo779

[131] Ridgwell, Andy and Daniela N. Schmidt. Past constraints on the vulnerability of marine calcifiers to massive carbon dioxide release, *Nature Geoscience,* 14 February 2010, 10.1038/ngeo755

[132] Merali, Zeeya. Climate blamed for mass extinction, *New Scientist*, April 1, 2006.

[133] Atkinson, Angus, et al. Long-term decline in krill stock an increase in salps within the Southern Ocean, Nature, 432, November 4, 2004, 100 -- 103.

[134] Proceedings of the National Academy of Sciences, DOI: 10.1073/pnas.0907998106.

[136] Nierenberg, Danielle, Ed. Rethinking the Global Meat Industry. State of the World: 2006. The WorldWatch Institute. London: Norton, 2006: 24.

[137] http://www.epa.gov/oecaagct/ag101/demographics.html

[138] Personal observation. The author has taught agribusiness courses for 35 years at Arizona State University.

[139] ERG, http://farm.ewg.org/farm/summary.php

[140] http://www.organicconsumers.org/articles/article_1371.cfm

[141] Food Planet, Food summit blames trade barriers, biofuels, June 4, 2008, www.planetark.com/dailynewsstory.cfm/newsid/48626/story.htm

[142] Edwards, Mark. Green Algae Strategy: End Biowar I and Engineer Sustainable Food and Biofuels, Tempe: LuLu Press, 2008, 40.

[143] Badger, Emily. We gotta save them. Miller-McCune.com, July 15, 2009.

[144] Worldwatch, 37.

[145] Yang, X. B. Soybean Brown Spot and Bacteria Blight. http://www.ipm.iastate.edu/ipm/icm/2003/7-28-2003/spotblight.html

[146] Why the Sudden Oak Death and Soybean Blight? http://www.jgi.doe.gov/sequencing/why/suddenoak.html

[147] MacKenzie, Debora, Caterpillar plague strikes West Africa, *New scientist*, 31 January 2009, 12.

[148] National Center for Integrated Pest Management, India. http://www.ncipm.org.in/asps/DisplayFertilizers.asp

[149] Bradsher, Keith and Martin, Andrew. Shortage and price of fertilizer threatens to make more hungry, Intl. Herald Tribune, May 1, 2008.

[150] FAO, ResourceSTAT, electronic database, at faostat.fao.org/site/405/default.aspx, updated 30 June 2007.

[151] Sharma, M. C. and Owen, L. A., J. Quat. Sci. Rev., 1996, 15, 335–365.

[152] MacKenzie, Debora. Melting glaciers will trigger food shortages, New Scientist, 20 March 2008.

[153] Sainath, P. The Largest Wave of Suicides in History, Counterpunch, February

12, 2009. http://www.counterpunch.org/sainath02122009.html

[154] Zwerdling, Daniel. India's Farming 'Revolution' Heading For Collapse, NPR, April 13, 2009. http://www.npr.org/templates/story/story.php?storyId=102893816

[155] Scherr, Sara and Sajal Sthapit. Farming and land use to cool the planet, in The State of the World 2009, the World Watch Institute, 2009, 37.

[156] Jones, A.J., R. Lal and D.R. Huggins. Soil erosion and productivity research: a regional approach. American Journal of Alternative Agriculture, 1997, 12:4, 183-192.

[157] Pimentel, David. Food, energy and society, 75.

[158] Young, A. Agroforestry for soil conservation. Wallingford, UK: CAB, 1989.

[159] Wilkinson, Bruce H. and Brandon J. McElroy, The impact of humans on continental erosion and sedimentation, *Geological Society of America Bulletin*, January 2007; v. 119; no. 1-2; p. 140-156; DOI: 10.1130/B25899.1

[160] WRI, World Resources Institute. New York: Oxford University Press, 1994.

[161] Faeth, P. and P. Crosson. Building the case for sustainable agriculture, Environment, 1994, 36:1, 16-20.

[162] Natural Resources Conservation Service, Soil Erosion, 2006.

[163] Pimentel, D., Hepperly, P., Hanson, J., Douds, D., and R. Seidel. 2005. Environmental, energetic, and economic comparisons of Organic and Conventional farming systems. Bioscience 55(7):573-582.

[164] Pimentel, David. Environment, Development and Sustainability. Dordrecht: Feb 2006. Vol. 8, Iss. 1, 119-137. USDA, Changes in average annual soil erosion by water on cropland and CRP land, 1992 -- 1997. Natural Resources Conservation Service. December, 2000.

[165] Sample, Ian. Global food crisis looms as climate change and population growth strip fertile land. The Guardian, 31 August 2007.

[166] Pimentel, David. Food, energy and society, 29.

[167] Lotter, D., Seidel, R., and W. Liebhardt. 2003. The performance of organic and conventional cropping systems in an extreme climate year. American Journal of Alternative Agriculture 18(2):1-9.

[168] Sundquist, Bruce. THE EARTH'S CARRYING CAPACITY ~SOME RELATED REVIEWS AND ANALYSES CHAPTER 6 ~ URBANIZATION-CAUSED TOPSOIL (CROPLAND) LOSS Edition 8 of June, 2009http://home.windstream.net/bsundquist1/se6.html

[169] Kijne, Jacob W. Unlocking the water potential of agriculture, Rome: FAO, 2003, 26.

[170] Sacks, W.J., B.I. Cook, N. Buenning, S. Levis, and J.H. Helkowski, 2009: Effects of global irrigation on the near-surface climate. *Clim. Dynam.*, **33**, 159-175, doi:10.1007/s00382-008-0445-z

[171] Gleick. The World's Water 2001. Washington, DC: Island Press, 2000: 52.

[172] Ibid., 52.

[173] Diouf, Jacques. *Turning the Tide Against Water Scarcity.* Food and Agriculture Organization of the United Nations, Mar. 2007. http://www.fao.org/english/dg/oped/index.html.

[174] Pearce, 24.

[175] Matt Weiser, Feds Document Shrinking San Joaquin Valley Aquifer Sacramento Bee, July13, 2009.

[176] California Department of Water Resources, 2009. http://www.owue.water.ca.gov/agdev/

[177] Gorman, Steve. California farms lose main water source to drought, Reuters, February 20, 2009.

[178] Burke, Garance. Will drilling more wells in California help or hurt?, The Associated Press, January 11, 2010.

[179] Ibid.

[180] Pollan, Michael.

[181] Pimentel, D. and Giampietro. The Tightening Conflict: Population, Energy Use, and the Ecology of Agriculture", 1994, http://dieoff.org/page69.htm

[182] International Energy Agency (IEA), Oil Market Report, Paris: August, 2001.

[183] Brown, J. and Foucher, S. http://graphoilogy.blogspot.com/

[184] USDA NASS, June 29, 2007 News Bulletin, http://www.nass.usda.gov

[185] FAO, Global fertilizer supply to outstrip demand, Feb 26, 2008. www.fao.org/newsroom/en/news/2008/1000792/index.html

[186] Galloway, J.N., J.D. Aber, J.W. Erisman, S.P. et. al. 2003. The nitrogen cascade. *Bioscience,* 53(4):341–356.

[187] Townsend, Alan R. and Robert W. Howarth. Fixing the Global Nitrogen Problem, Scientific American, Feb 2010, 64-73.

[188] Gurian-Sherman ,Doug and Noel Gurwick. No Sure Fix: Prospects for Reducing Nitrogen Fertilizer Pollution through Genetic Engineering. Union of Concerned Scientists, December 2009.

[189] Krupa, S.V. 2003. Effects of atmospheric ammonia, (NH3) on terrestrial vegetation: A review. *Environmental Pollution* 124:179–221.

[190] Rockström, J., W. Steffen, K. Noone, Å. Persson,F.S. Chapin et. al. 2009. A safe operating space for humanity. *Nature,* 461:472–475.

[191] National Institutes of Health. Air Pollution & Respiratory Disease, July, 2009.

---

[192] Keim, Brandon. Reactive Nitrogen: The Next Big Pollution Problem, Wired Science, May 15, 2008.

[193] DNR, "State of Iowa, Public Drinking Water Program 2006 Annual Compliance Report" (Des Moines, IA: June, 2007.

[194] Pimentel, D., Hepperly, P., Hanson, J., Douds, D., and R. Seidel. 2005. Environmental, energetic, and economic comparisons of Organic and Conventional farming systems. Bioscience 55(7):573-582.

[195] Vaccari, David. Phosphorus Famine: The Threat to Our Food Supply, *Scientific American*, June, 2009.

[196] Saber K, Nahla L, Ahmed D, Chedly A. 2005. Effect ofP on nodule formation and N fixation in bean. Agron Sustainable Develop. 25:389-393.

[197] Rodriguez H, Fraga R. 1999. Phosphate solubilizing bacteria and their role in plant growth promotion. Biotechnol Adv. 17:319-339

[198] Gyaneshwar P, Naresh KG, Parekh LJ, Poole PS. 2002. Role of soil microorganisms in improving P nutrition of plants. Plant Soil. 245:83-93.

[199] Chandini TM, Dennis P. 2002. Microbial activity, nutrient dynamics and litter decomposition in a Canadian Rocky Mountain pine forest as affected by N and P fertilizers. Forest Ecol Manage. 159: 187-20l.

[200] Lewis, Leo. Scientists warn of lack of vital phosphorus as biofuels raise demand, The Times, June 23, 2008.

[201] Murphy, Annie. Bolivian farmers fueled up by soybeans, marketplace, American public radio, January 27, 2009.

[202] Edwards, Mark. Biowar I, 2007, 78.

[203] Ibid.

[204] Clay, Jason. Speech to the Global Institute of Sustainability, ASU, World Wildlife Fund., January 20, 2010

[205] UTEX Algae Collection, http://www.utex.org/

[206] Ugwu CU, Aoyagi H, Uchiyama H. Photobioreactors for mass cultivation of algae. *Bioresource Technology*, 2008, 99(10):4021-8.

[207] Florida Department of Environmental Protection, www.bioreactor.org/

[208] Schreiner RP, Mihara KL, McDaniel H, Bethlenfalvay GJ. 1997. Mycorrhizal fungi influence plant and soil functions and interactions. Plant Soil, 188:199-209.America, Elsevier Academic Press, 2005.

[210] U.S. DOE, 2009. National Algal Biofuel Technology Roadmap. https://e-center.doe.gov/iips/faopor.nsf/UNID/79E3ABCACC9AC14A852575CA00799D99/$file/AlgalBiofuels_Roadmap_7.pdf

[211] Edwards, Mark R. Algal Industry Survey, Algal Biomass Organization Webinar, Jan 14, 2010. http://www.algalbiomass.org/

[212] Galloway, J. N. et al., Science 320, 889 (2008).

[213] Conway, Gordon. The doubly green revolution: food for all of the 21st century, Ithaca: Cornell University press, 1997, 47.

[214] Goyal, SK. Algal Biofertilizers, In Biotechnology of biofertilizers, in Kannaiyan, S. Ed., Delhi: Narosa Publishing House, 2002, 4.

[215] Baratharkur, HB and H. Talukdar. Use of azolla and commercial mito of algal biofertilizers gen fertilization in India, international Rice research newsletter, 1983, 8:2, 20.

[216] Bagyaraj, DJ and a K. Varma. State of art of mycorrhizae in mycorrhiza roundtable. Proceedings workshop held in New Delhi, IDRC, New Delhi India, 1988, 25 – 32.

[217] Rao, Subba. An appraisal of biofertilizers in India, in Biotechnology of biofertilizers, in Kannaiyan, Ed., Delhi: Narosa Publishing House, 2002, 4.

[218] Venkataraman, G.S. the role of blue-green algae in tropical rice cultivation. In nitrogen fixation by free living microorganisms, Stewart, W.D.P. (ed.), U.K.: Cambridge University Press, 1976, 207 – 218.

[219] Rao Sandra and Sinha, M.K. Phosphate dissolving organisms in soil. Indian Journal of agricultural science, 1963, 33:272 -- 278.

[220] Ibid. 275.

[221] Ibid., 273.

[222] Powell, CL and DJ Bagyaraj, (eds). VA Mycorrhiza. CRC Press, Florida, 1976.

[223] Mosse, B. and PB Tinker, (eds) Endomycorrhizas. London: Academic press, 1984.

[224] Wani, SP and KK Lee. Biofertilizers for sustainable cereal crop production, in Biotechnology of biofertilizers, Kannaiyan, S. Ed., Delhi: Narosa Publishing House, 2002, 50-64.

[225] McGovern, George. Ending Hunger Now, Fortress Press, 2005, 23.

[226] Edwards, Mark. BioWar I: Why Battles over Food and Fuel Lead to World Hunger, LuLu Press, 2007, 84.

[227] Pimentel, David. Food, energy and society, 73.

[228] Dias De Oliveira, Marcelo E., Burton E. Vaughan, and Edward J. Rykiel Jr. "Ethanol as Fuel: Energy, Carbon Dioxide Balances, and Ecological Footprint." *BioScience 55:7 (2005):* 593–602.

[229] *Linking Land Quality, Agricultural Productivity, and Food Security.* United States Department of Agriculture, Economic Research Service AER-823, 31. http://www.ers.usda.gov/publications/aer823/aer823e.pdf.

[230] http://www.organicconsumers.org/articles/article_1371.cfm

[231] Food Planet, Food summit blames trade barriers, biofuels, June 4, 2008, www.planetark.com/dailynewsstory.cfm/newsid/48626/story.htm

[232] Sachs , Jeffrey D. *The End of Poverty: Economic Possibilities for Our Time,* Penguin Press, 2005.

[233] Edwards, Mark R. *Green Solar Gardens: Algae's Promise to end Hunger,* Tempe: CreateSpace, 2009, 22.

[234] USDA NASS, June 29, 2007 News Bulletin, http://www.nass.usda.gov

[235] Ameziane, R., K. Bernhard, and D. Lightfoot. 2000. Expression of the bacterial *gdhA* gene encoding a NADPH glutamate dehydrogenase in tobacco affects plant growth and development. *Plant and Soil* 221:47–57.

[236] Weiss, Rick. Firms Seek Patents on 'Climate Ready' Altered Crops, Tuesday, May 13, 2008; Page A04ps, Washington Post, May 13, 2008; A4.

[237] Marrero, Carmelo Ruiz. Biotech Bets on Agrofuels, Center for International Policy (CIP), April 24, 2008. http://americas.irc-online.org/am/5179

[238] Bloomberg News, Antitrust Questions for Monsanto, Jan 14, 2010.

[239] Organic consumers' organization. Monsanto's GE Seeds are Pushing US Agriculture into Bankruptcy, http://www.organicconsumers.org/monlink.cfm

[240] Union of Concerned Scientists. Bush administration, biotech industry, and agribusiness overstate genetically engineered crops' potential to solve world food crisis, June 26, 2008.

[241] Benbrook, Charles. The impacts of genetically engineered crops: the first 13 years, Union of Concerned Scientists, November, 2009.

[242] Participant Media and Karl Weber. Food Inc.: How Industrial Food is Making Us Sicker, Fatter, and Poorer-and What You can do about It, Public Affairs, 2009.

[243] Berry, Wendell and Michael Pollan Bringing It to the Table: On Farming and Food , Counterpoint, 2009.

[244] Holmes, J., The Washington Times op-ed column, Oct. 1, 2008.

[245] Hillary Clinton, Attacking hunger at its roots, *The Huffington Post*, June 11, 2009.

[246] http://www.worldwatch.org/node/810

[247] Adapted from: Sullivan, Preston. Appropriate Technology Transfer for Rural Areas, (ATTRA). http://www.soilandhealth.org/01aglibrary/010117attrasoilmanual/

[248] Edwards, Clive A. and P.J. Bohlen. 1996. Biology and Ecology of Earthworms. Chapman and Hall, New York. 426.